Chile Earthquake of 2010

Lifeline Performance

PREPARED BY

Earthquake Investigation Committee of the Technical Council of Lifeline Earthquake Engineering

EDITED BY

Alex K. Tang, P.E., P.Eng., C.Eng., F.ASCE
John M. Eidinger, P.E., M.ASCE

SPONSORED BY

Technical Council on Lifeline Earthquake Engineering
Monograph No. 36

ASCE

Published by the American Society of Civil Engineers

Library of Congress Cataloging-in-Publication Data

American Society of Civil Engineers. Earthquake Investigations Committee.
Chile earthquake of 2010 lifeline performance / prepared by Earthquake Investigation Committee of the Technical Council of Lifeline Earthquake Engineering ; edited by Alex K. Tang, P.E., P. Eng., C. Eng., F. ASCE, John M. Eidinger, P.E., M. ASCE ; sponsored by Technical Council on Lifeline Earthquake Engineering.
pages cm. -- (Monograph ; no. 36)
Includes bibliographical references and index.
ISBN 978-0-7844-1282-4 (print) -- ISBN 978-0-7844-7784-7 (pdf)
1. Lifeline earthquake engineering--Chile--Evaluation. 2. Earthquake damage--Chile. 3. Chile Earthquake, Chile, 2010 (February 27) I. Tang, Alex. II. Eidinger, John M. III. Title.
TA654.6.A434 2013
624.1'762--dc23

2013020231

Published by American Society of Civil Engineers
1801 Alexander Bell Drive
Reston, Virginia 20191
www.asce.org/pubs

ISBN 978-0-7844-1282-4 (print)
ISBN 978-0-7844-7784-7 (PDF)
Manufactured in the United States of America.

TCLEE Monograph Series

These publications may be purchased from ASCE via telephone at 1-800-548-ASCE (2723) or online at www.asce.org/bookstore. The TCLEE Web site is www.asce.org/tactclee.

No. 1 Recent Lifeline Seismic Risk Studies, Kiremidjian, Anne S., Editor, 1990.

No. 2 Seismic Lost Estimates for a Hypothetical Water System, A Demonstration Project, Taylor, Craig E., Editor, August 1991.

No. 3 Guide to Post-Earthquake Investigations of Lifelines, Schiff, Anshel J., Editor, August 1991.

No. 4 Lifeline Earthquake Engineering, Proceedings of the 3rd U.S. Conference on Lifeline Earthquake Engineering, August 22-23, 1991, Los Angeles, CA, Cassaro, Michael, Editor, August 1991.

No. 5 Lifeline Earthquake Engineering in the Central and Eastern United States, Ballantyne, Donald, Editor, September 1992.

No. 6 Lifeline Earthquake Engineering, Proceeding of the 4th U.S. Conference on Lifeline Earthquake Engineering, August 10-12, 1995, San Francisco, CA, O'Rourke, Michael J., Editor, August 1995.

No. 7 Critical Issues and State of the Art on Lifeline Earthquake Engineering, Schiff, Anshel J. and Buckle, Ian, Editors, October 1995.

No. 8 Northridge Earthquake: Lifeline Performance and Post-Earthquake Response, Schiff, Anshel J., Editor, August 1995.

No. 9 Seismic Design for Natural Gas Distributors, McDonough, Peter W., August 1995.

No. 10 Methods of Achieving Improved Seismic Performance of Communications Systems, Tang, Alex, and Schiff, Anshel J., Editors, September 1996.

No. 11 Guide to Post-Earthquake Investigation of Lifelines, Schiff, Anshel J., Editor, July 1997.

No. 12 Seismic Guidelines for Ports, Werner, Stuart D., Editor, March 1998.

No. 13 Overcoming Barriers: Lifeline Seismic Improvement Programs, Taylor, Craig E., Mittler, Elliott, and Lund, Le Val, September 1998.

No. 14 Hyogo-Ken Nambu Earthquake of January 17, 1995—Lifeline Performance, Schiff, Anshel J. Editor, 1998.

No. 15 Guidelines for the Seismic Evaluation and Upgrade of Water Transmission Facilities, Eidinger, John M. and Avila, Ernesto A., Editors, January 1999.

No. 16 Optimizing Post-Earthquake Lifeline System Reliability (Proceedings of the 5th U.S. Conference on Lifeline Earthquake Engineering, Seattle, Washington, August 12-14, 1999.), Elliott, William M., and McDonough, Peter W., Editors, August 1999.

No. 17 Ismit (Kocaeli), Turkey Earthquake of August 16, 1999, Including Duzce Earthquake of November 12, 1999—Lifeline Performance, Tang, Alex K., Editor, September 2000.

No. 18 Chi-Chi, Taiwan, Earthquake of September 21, 1999—Lifeline Performance, Schiff, Anshel J., and Tang, Alex K., Editors, October 2000.

No. 19 Gujarat (Kutch) India, M7.7 Earthquake of January 26, 2001 and NAPA M5.2 Earthquake of September 3, 2000, Eidinger, John M., Editor, June 2001.

No. 20 The Nisqually, Washington, Earthquake of February 2001—Lifeline Performance, McDonough, Peter W., Editor, February 2002.

No. 21 Acceptable Risk Process—Lifelines and Natural Hazards, Taylor, Craig E., and VanMarcke, Erik H., Editors, March 2002.

No. 22 Seismic Screening Checklists for Water and Wastewater Facilities, Heubach, William F., Editor, September 2002.

No. 23 Atico, Peru Mw 8.4 Earthquake of June 23, 2001, Curtis L. Edwards, Editor, October 2002.

No. 24 Lifeline Performance of El Salvador Earthquakes of January 13 and February 13, 2001, Le Val Lund, Editor and Carl Sepponen, Editor, September 2002.

No. 25 Advancing Mitigation Technologies and Disaster Response for Lifeline System: Proceedings of the Sixth U.S. Conference and Workshop on Lifeline Earthquake Engineering, Beavers, J.E., Editor, August 2003.

No. 26 Fire Following Earthquake, Scawthorn, Charles, Eidinger, John M., and Schiff, Anshel J., Editors, 2005.

No. 27 Zemmouri, Algeria, Mw 6.8 Earthquake of May 31, 2003, Edwards, Curtis L., Editor, 2004.

No. 28 San Simeon Earthquake of December 22, 2003 and Denali, Alaska, Earthquake of November 3, 2002, Yashinsky, Mark, Editor, Lund, Le Val, Co Editor, 2004.

No. 29 Hurricane Katrina: Performance of Transportation Systems, DesRoches, Reginald, Editor, 2006.

No. 30 Sumatra-Andaman Islands Earthquake and Tsunami of December 26, 2004 Lifeline Performance, Strand, Carl and Masek, John, Editors, 2007.

No. 31 Kashiwazaki, Japan Earthquake of July 16, 2007 Lifeline Performance, Tang, Alex K.K. and Schiff, Anshel J., Editors, 2007.

No. 32 Pisco, Peru Earthquake of August 15, 2007 Lifeline Performance, Tang, Alex K.K. and Johansson, Jorgen, Editors, 2007.

No. 33 Pacific Northwest Storms of December 1-4, 2007: Lifeline Performance, edited by Teresa Elliott, P.E., and Alex K. Tang, P.E.

No. 34 Seismic Resilience of Natural Gas Systems: Improving Performance, edited by Peter W. McDonough, P.E.

No. 35 Haiti M_w 7.0 Earthquake of January 12, 2010: Lifeline Performance, edited by Curtis L. Edwards, P.E.

No. 36 Chile Earthquake of 2010: Performance of Lifelines, edited by Alex K. Tang and John Eidinger.

Other TCLEE Publications

Duke, C. Martin, Editor, The Current State of Knowledge of Lifeline Earthquake Engineering, Proceedings TCLEE Specialty Conference, August 30-31, 1977, Los Angeles, CA, (Later designated as the 1st U.S. Conference on Lifeline Earthquake Engineering), August, 1977.

Dowd, Munson, Editor, Annotated Bibliography on Lifeline Earthquake Engineering, 1980.

Smith, D. J, Jr., Editor, Lifeline Earthquake Engineering, The Current State of Knowledge 1981, Proceedings of the Second TCLEE Specialty Conference, August 20-21, 1981, Oakland, CA, (Later designated as the 2nd U.S. Conference on Lifeline Earthquake Engineering), August 1981.

Hall, William J., Advisory Notes on Lifeline Earthquake Engineering, 1983.

Nyman, Douglas, NSF Principal Investigator, Guidelines for the Seismic Design of Oil and Gas Pipelines Systems, TCLEE Committee on Gas and Liquid Fuels, 1984.

Cooper, James, Editor, Lifeline Earthquake Engineering Performance, Design and Construction, 1984.

Cassaro, Michael and Martinez-Romero, E., Editors, The Mexico Earthquake, 1985, Factors Involved and Lessons Learned, 1986.

Eguchi, Ronald and Crouse, C. B., Lifeline Seismic Risk Analysis - Case Studies, 1986.

Wang, Leon R. L. and Whitman, Robert, Seismic Evaluation of Lifeline Systems-Case Studies, 1986.

Cassaro, Michael and Cooper, James, Editors, Seismic Design and Construction of Complex Civil Engineering Systems, 1988.

Werner, Stuart D. and Dickenson, Stephen E., Editors, Hyogo-Ken Nambu (Kobe) Earthquake of January 17, 1995: A Post-Earthquake Reconnaissance of Port Facilities, TCLEE Ports Committee, 1996.

ASCE Manual

Schiff, Anshel J., Editor, Guide to Improved Earthquake Performance of Electric Power Systems, ASCE Manual 96.

TCLEE Earthquake Investigation Reports

TCLEE has also prepared numerous earthquake reports that have appeared in other publications. References to these reports and 10 short reports associated with TCLEE monographs can be viewed on the ASCE/TCLEE Web site address given below. The 10 short reports are each about 5 to 15 pages long, contain a summary of main observations and some pictures. They can be downloaded at *www.asce.org/tactclee.*

Contents

Preface

The Earthquake Investigation Committee of the Technical Council of Lifeline Earthquake Engineering (TCLEE), American Society of Civil Engineers (ASCE), was established to initiate, organize, train for, coordinate, and evaluate the performance of lifelines following earthquakes. Members of the committee are employees of lifeline industries, consulting engineers, and academics from the United States and Canada. Committee members provide services on a voluntary basis. For some earthquake investigations, companies of participants do not require an individual to take vacation time for the investigation and may provide some support for expenses. ASCE also provides support to reimburse expenses. In addition to the time associated with the reconnaissance trip, the substantial effort by each individual to prepare a short report for the TCLEE Web page and the full report for the monograph series is all done on a voluntary basis. The cost of this effort is substantially more than the support provided by ASCE.

While individuals participating in the investigation do not need to be members of the committee or ASCE, they are expected to follow the committee's earthquake investigation practices as described in the ASCE publication, TCLEE Monograph 11, *Guide to Post-Earthquake Investigation of Lifelines*. Members of the investigation team coordinate with other groups and may participate in groups organized by other organizations. They gather data from both good and poor lifeline performance from domestic and foreign earthquakes to provide information for practitioners to improve the performance of the lifeline systems. Foreign earthquakes that have been investigated by TCLEE teams include the 1985 Chile, 1988 Soviet Armenia, 1990 Philippines, 1991 Costa Rica, 1992 Turkey, 1994 Kobe (Japan), 1999 Kocaeli (Turkey), 1999 Chi-Chi (Taiwan), 2001 Gujarat (India), 2002 Atico (Peru), 2004 Zemmouri (Algeria), 2004 Sumatra, 2007 Kashiwazaki (Japan), 2008 Pisco (Peru) earthquakes, 2008 Wenchuan (China) earthquake, and 2009 L'Aquila (Italy) earthquakes. Domestic earthquakes that have been investigated include the 1989 Loma Prieta, 1992 Landers, 1994 Northridge, 2000 Napa, 2001 Nisqually, 2002 Denali, 2003 Paso Robles, and 2008 Alum Rock earthquakes. TCLEE also started to document major storm impact on lifelines: The Pacific Northwest Storm of December 2007 was our first post windstorm investigation.

The Kobe earthquake report, Monograph 14, was the first foreign earthquake investigation report published by ASCE as a TCLEE monograph. The first domestic earthquake investigation report published by ASCE as a TCLEE monograph was Number 8, the Northridge earthquake. Before this, TCLEE lifeline reports were published by the Earthquake Engineering Research Institute (EERI). The Earthquake Investigation Committee continues to cooperate with EERI to provide an abbreviated version of lifeline performance reports in Earthquake Spectra, an EERI publication. TCLEE also publishes brief preliminary reports on the ASCE/TCLEE Web page.

Alex Tang, PE, F. ASCE
John Eidinger, PE, M. ASCE

Contributors

Name	Affiliation	E-mail
Alex Tang	L&T Consultant	alexktang@mac.com
John Eidinger	G&E Engineering System	eidinger@geengineeringsystems.com
Tom Cooper	T.W. Cooper Inc.	twcoopinc@aol.com
Leonardo Dueñas-Osorio	Rice University	alvis@tdmrc.org
Bill Fullerton	The Louis Berger Group Inc.	wfullerton@louisberger.com
Roy Imbsen	Earthquake Protection Systems, Inc.	roy.imbsen@earthquakeprotection.com
Leon Kempner	Bonneville Power Administration	lkempnerjr@bpa.gov
Alexis Kwasinski	University of Texas, Austin	akwasins@mail.utexas.edu
Allison Pyrch	Shannon & Wilson Inc.	AMP@shanwil.com
Anshel J. Schiff	Precision Measurement Instruments	schiff@stanford.edu
Yumei Wang	Oregon Department of Geology and Mineral Industries	yumei.wang@dogami.state.or.us

Acknowledgments

Most of the contributing authors of this report are members of the Earthquake Investigation Committee (EIC) of Technical Council on Lifeline Earthquake Engineering (TCLEE). Anshel Schiff, Yumei Wang, and Alex Tang were fully funded, while Allison Pyrch, Tom Cooper, and Leonardo Dueñas-Osorio were partially funded by ASCE. ASCE support was received at the direction of Pat Natale and the ASCE executive directors, with unparalleled attention from John Durrant, John Segna, and Tenzing Barshee to help speed up the process. A unanimous decision to approve the trip was granted by ASCE Executive Committee. The investigation team is grateful to many local support persons who provided either contacts or approvals to visit lifeline facilities, collecting relevant lifelines performance information as well as furnishing perishable data and photos for this monograph. We would also like to thank Allison Pyrch and Yumei Wang, who set up meetings and arrange vehicle rentals in Santiago two days prior to the team's arrival. Many team members worked hard prior to the trip to align local contacts for access to secured facilities, particularly Prof. Alexis Kwasinski, who worked extra hours to line up contacts in the Department of Public Works in Santiago. The trip would not have been successful without the four important Spanish translators—Allison Pyrch, Prof. Leonardo Dueñas-Osorio, Prof. Alexis Kwasinski, and Bill Fullerton.

As is not uncommon in post-earthquake reconnaissance, incomplete information in the weeks and months following an event can lead to omissions and misunderstandings. We apologize if the findings in this report are incomplete; the reader is cautioned that it may take months to years of post-earthquake evaluations before a comprehensive understanding of lifeline impacts is available. ASCE/TCLEE will continue to update the lifeline performance information as additional information is collected.

The findings and photos presented in this report reflect the collected input from many people and sources, all of whom are credited wherever the information or the photograph appears in this document. We are thankful to them providing us with permission to use the information and photos. The ASCE TCLEE team members, who are the authors of this monograph, collected data and took photos from April 9 to 17, 2010.

Alex K. Tang, P.E., P. Eng., C.Eng., F.ASCE,
John Eidinger, P.E., M.ASCE

Acknowledgment of Local Support

We are grateful to the many individuals and organizations (companies, governmental organizations, and universities) that provided the team members with assistance and support before, during, and after the post investigation effort. A number of individuals requested that their names not be included in the monograph; we are deeply in debt to their support and information given.

We are also much in debt to Professor Mauricio Villagran, who accompanied us for two full days providing an overview of damage around Concepción and helping with many local lifeline facility key contacts to make our investigation more complete. He also set up an information sharing session on campus for the team.

Academic

Prof. Mauricio Villagran of Universidad Católica de la Santísima Concepción

Mr. Hugh Rudnick of Catholic University, Santiago

Mr. Solange Loyer of Universidad Católica de la Santísima Concepción

Mr. Christian Ledezma of Pontificia Universidad Catolica de Chile

Mr. Roberto J. Mora Mella, Rector, Colegio Concepción San Pedro

Company & Consultant

Ms. Sonia Castellanos of Emerson Network Power

Mr. Juan Pablo Muñoz Díaz of LAN Chile Airlines

Mr. Nicolas Billikopf Mujica of LAN Chile Airline

Ms. Karla Gonzalez of GHD Consulting

Mr. Geri Pranzini of GHD Consulting

Mr. Miguel Valenzuela of GHD Consulting

Mr. Juan Pablo Gardeweg of Editorial Compass

Ms. Barbara Domont, friend of Allison Pyrch

Mr. Nicolás Lira Turpaud, friend of Allison Pyrch

Geotechnical

Mr. Manuel Suarez of Servicios Nacional de Geologia y Minerales, Geology Director for Sernageomin

Mr. Jose Antonio Naranjo of Servicios Nacional de Geologia y Minerales

Mr. Javier Fernandez of Servicios Nacional de Geologia y Minerales

Ms. Mariana Cervetto, Geologist

Government

Mr. Luis Estelle Aguirre, Government of Chile, Santiago

Mr. Carlos Sanhueza Sanchez, Government of Chile, Concepción

Electric Power

Mr. Eduardo Ricke from the CDEC-SIC (Centro de despacho economico de carga - Sistema interconectado central)

Mr. Eduardo Andrade of Transelec

Mr. Claudio Pizarro of Transelec

Mr. Ramon Vazquez of Transelec

Mr. Luis Perez of Transelec

Mr. Eduardo Zambrano of Transelec

Mr. Mario Donoso, General Manager of CGE

Mr. Gonzalo Fillol of CGE

Mr. Javier Herrera of CGE

Mr. Jaime Bertin of CGE Transmission

Mr. Jose Castro of CGE Distribution

Mr. Javier Herrera of CGE Transmission

Mr. Luis Martinez Vasquez of CGE Transmission

Mr. Manuel Ortega of CGE Distribution

Mr. Juan Rosales of CGE Distribution

Mr. Sergio Trabolt of CGE Distribution

Ports (Seaport and Airport)

Mr. Eduardo Mesina of Ministry of Public Works - Director of Ports

Mr. Marco Araneda of Minstry of Public Works – Ports Director Region VIII

Ms. Carolina Gutierrez of Ministry of Public Works – Acting Director, Ports Projects

Mr. Pedro Gonzalez of Lirquen Port

Mr. Eduardo Faunez of Coronel Port

Mr. Guillermo Guzman Solar of Ministry of Public Works – Director, Airports Projects

Mr. Eduardo del Canto Hidalgo of Civil Aeronautics Administration

Mr. Rodrigo Lopez Muela Parra of Civil Aeronautics Administration

Mr. Hector Sandoval of Civil Aeronautics Administration (DGAC) – Concepción Airport Acting Administrator

Mr. Rodrigo Esquivet Opazo of Asesoria Inspección Fiscal Aeropuerto Carriel Sur –

Concepción Airport

Ms. Alejandra Ruiz of Civil Aeronautics Administration (DGAC) – Los Angeles Airport Administrator

Gas & Liquid Fuel

Mr. Fernando Raiqueo Vaccaro of Enap Refineries

Telecommunications

Mr. Carlos Urzua of Telefonica,

Ms. Francisca Vilches of ENTEL, Concepción

Mr. Victor Morales of ENTEL, Concepción

Mr. Luis Leal of ENTEL, Concepción

Transportation

Mr. Juan Arrese Luco of Ministry of Public Works - Director of Hydraulics

Mr. Luis Estelle Aguirre, Ministry of Public Works – Hydraulics

Mr. Alejandro Kolbe of Ministry of Public Works

Mr. Jorge Campusano, Ministry of Public Works - Bridge Constructión

Water & Wastewater

Mr. Jose Luis Arrano Urzua of Essbio

Mr. Claudio Santelices Boettcher of Essbio

Allison was trying to find the place of our meeting at the Chile Geological Department on April 8, 2010. The information was in her laptop, and Roy became the mobile table.

After the presentations concluded, we entered into a friendly discussion on how to improve the process of earthquake mitigation particularly in the lifelines sector. On the left, Manuel Suarez, Geology Director for Sernageomin, is the host of the meeting.

Team members took the Santiago Metro back to the hotel after the meetings. Roy Ibmsen is just one step behind Allison Pyrch.

Our first group dinner on April 8, 2010. Left to right, Tom Cooper, Anshel Schiff, John Eidinger, Alexis Kwasinski, Roy Imbsen, Yumei Wang, Allison Pyrch, and Leon Kempner. Not photographed are Bill Fullerton and Leonardo Dueñas-Osorio, who arrived on April 9 and 10, respectively

On April 11, Prof. Mauricio Villagran (first person on the right) led us through the security to this tsunami devastated fish factory on Isla Rocuant in Talcahuano.

A group photo after a successful meeting with the Minister of Public Works in Santiago prior to our trip to Concepcion. From left to right, Geri Pranzini, Miguel Valenzuela, Tom Cooper, Roy Imbsen, Karla Gonzalez, Yumei Wang, Allison Pyrch, and John Eidinger.

Even after a long day of work, team members still have a smile for the camera. From left to right seated: John Eidinger, Leon Kempner, and Anshel Schiff. From left to right standing: Roy Imbsen, Leonardo Dueñas-Osorio, Yumei Wang, Allison Pyrch, Bill Fullerton, Alexis Kwasinski, Alex Tang, and Tom Cooper.

Before the boat ride, a group photo was taken by Alexis Kwasinski. From left to right: Alex Tang, Allison Pyrch, Coronel Port staff engineer, Mauricio Villagran, Roy Imbsen, and Eduardo Faúndez R. of Coronel Port.

The last team photo in Santiago taken on April 17 before seven team members headed home. From left to right, John Eidinger, Anshel Schiff, Roy Imbsen, Leon Kempner, Yumei Wang, Allison Pyrch, Alex Tang, Alexis Kwasinski, and Leonardo Dueñas-Osorio. Both Roy and Alex stayed behind. Not pictured, Bill Fullerton.

1 INTRODUCTION

On February 27, 2010 at 3:34 a.m. local time, a magnitude (M_w) 8.8 earthquake hit the west coast of the Maule Region of Chile. The epicenter of this subduction earthquake was located offshore at 35.909°S, 72.733°W with a depth of 35 km, about 330 km SW of Santiago and 105 km NNE of Concepción. Concepción is more than 500 km south of the Chilean capital of Santiago. Figure 1.1 shows the epicenter's location and the distance between Santiago and Concepción. The rupture zone was about 150 km by 500 km (Fig. 1.2). There were more than 200 aftershocks of magnitude 6 or higher during the first week after the main event. The total loss was estimated at about $30 billion. It will take months to consolidate the real loss figures as the estimate was based on the temporary repairs, fixes, and social relief needed to bring some normality to the earthquake-impacted areas.

Because this area is known for large earthquakes of magnitude 8 or higher, the Chilean government encourages earthquake mitigation in structures and infrastructures. This earthquake demonstrated the positive effects of these efforts. Both fatality and damage were small for the magnitude of this earthquake.

The overall lifelines performance was moderate to good. This earthquake removed all vulnerable items that were not upgraded or installed with best practices. A brief description of the performance of each lifeline is presented in this chapter; more detailed descriptions are provided in the following chapters.

The tsunami that followed the earthquake did more damage than the earthquake itself.

1.1 Geotechnical Aspects

The causes of lifelines damage and service disruptions were poor soil conditions, lateral spreading, and strong shaking. It was evident that buildings of similar structural design and construction in the same area were damaged and some on treated soil were not damaged. The buildings on treated foundations sustained minor non-structural damage. Because the earthquake occurred at the end of summer and the dry season there were only a few landslides with minimal impact to lifelines.

The tsunami that followed the earthquake was responsible for half the losses—from both deaths and property loss.

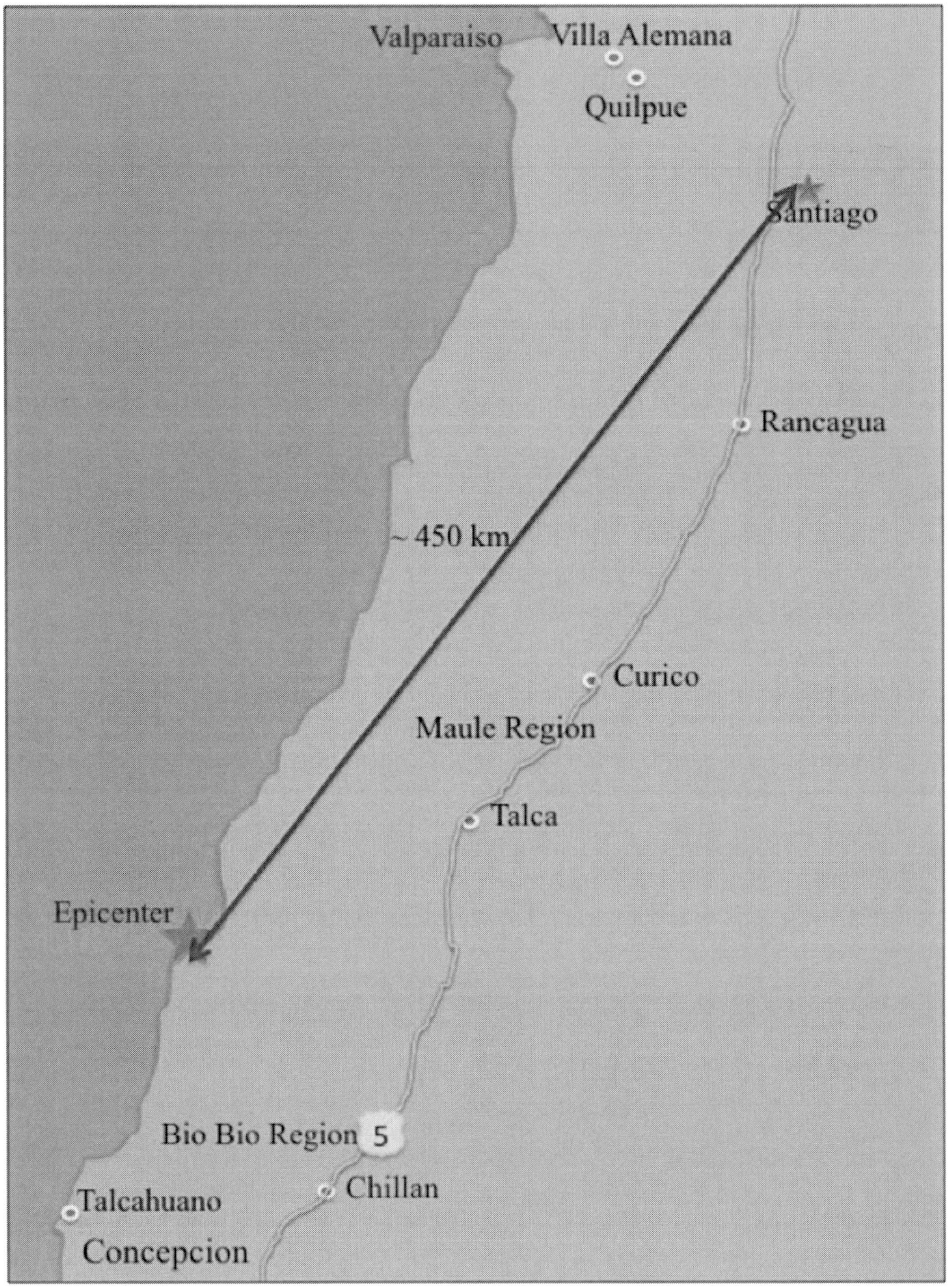

Figure 1.1. The location of the epicenter relative to Concepción and Santiago. The hardest hit areas are the Maule and Bio Bio regions.

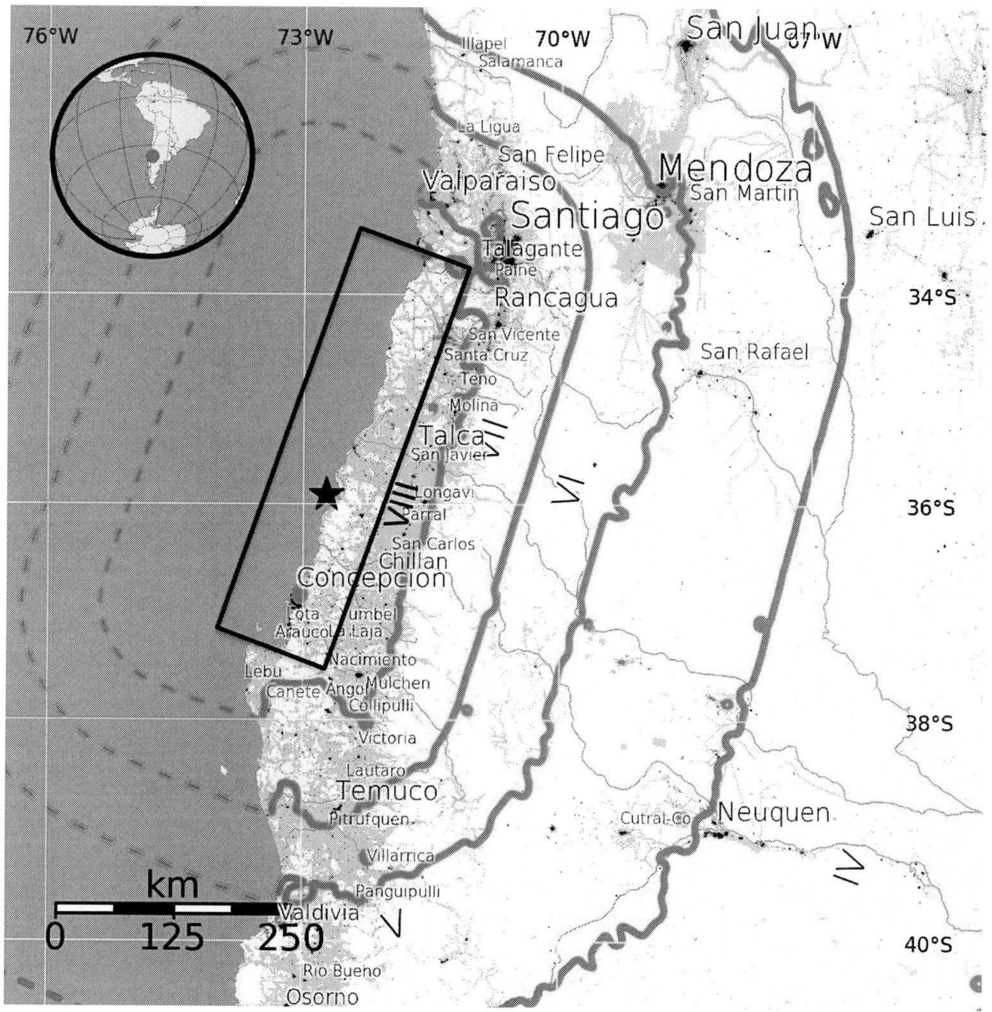

Figure 1.2. Contours of the seismic exposure based on MMI scale. The rectangle indicates the rupture zone of this large seismic event. (Source: USGS)

1.2 Lifelines

Power Systems

The electric power transmission and distribution networks in the earthquake-impacted areas sustained moderate damage. The transmission network was ready to energize within 24 hours of the event. The distribution system took from 2 to 4 days to restore service mostly due to building damage. In a few areas, the outage lasted 5 or 6 days as the power companies had problems identifying power outage locations. Two weeks after the earthquake, there was a 2-hour power outage (see Chapter 6). The modes of failure in the distribution system were toppled utility poles, pole mounted transformers falling or sustaining internal damage, and drop wires that were disconnected due to partial building failure or utility pole movement.

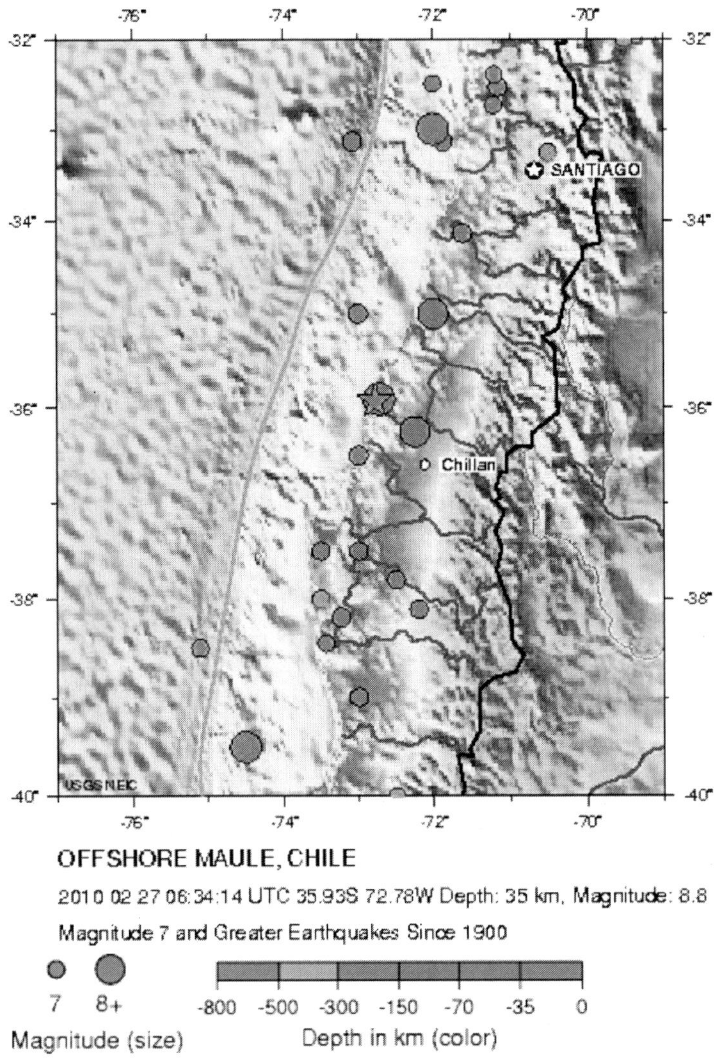

Figure 1.3. Earthquakes of magnitude 7 and higher since 1900 in the region. Note that most of the Magnitude 8+ events occurred along the coast of Chile (Source: USGS)

The power outage caused the cellular network malfunction that hindered emergency response and rescue immediately after the earthquake.

Communications

Communications systems, particularly cellular networks, performed poorly. Cellular phones are more commonly used in the region. All cellular service providers restored service within 7 days. The communications disruption hindered restoration of the water and power distribution systems, which depend on cellular phone as their service department communication tool. The main reason for the cellular service failure was loss of power. A number of cellular antennas were damaged, batteries fell off the rack, and equipment was damaged. Most of the standalone cellular antenna towers performed well; the damage antennas were those installed on building roofs and water towers. The landline performed moderately well, only losing service for about 3 days. Fiber optic cable that was co-located with bridges was severed due to bridge collapse.

Water Systems

The water system sustained heavy damage particularly in Concepción. There were more than 3000 pipe breaks and damage to the purification facilities. The looting and violence in the Concepción area and the cellular phone service failure delayed repairs to pipes and pumping facilities.

One fire following the earthquake occurred in the chemical building of the University of Concepción. The building was almost gutted, and much research data was lost.

Transportations (Highways, Bridges, and Railway)

The road system sustained extensive damage due to collapsed bridges and overpasses. Road failures were caused by upheaval, cracks, and lateral spreading of road surfaces. As the major road system was privatized, the service restoration was done within a few days using detours and bailey bridges.

The train system in some areas was still out of service during the TCLEE investigation 6 weeks after the earthquake. The main reasons were damaged power lines and a few damaged terminal buildings that were not repaired. Within the city of Concepción the commuter train was operating.

Sea Ports and Airports

Both the Santiago and Concepción airports sustained damage to the terminal buildings and control towers; however, the runways were not damaged. Within days of the event, Santiago Airport was back to service with a limited number of flights using temporary tents to process passengers. Concepción Airport opened within a day due to lighter damage. When the communication system between these two airports came back on line after 24 hours, many flights were redirected to Concepción Airport.

Most of the ports within the Concepción area sustained various degrees of damage. There was no damage to the piers at the Lirquen Port; however, the warehouses in the port sustained extensive damage, and two small warehouses collapsed. The new seismically isolated pier at Coronel Port performed well, while the other pier at this port sustained significant damage, particularly the portion close to shore.

Gas and Tanks (and Refinery)

The refinery in Concepción was closed due to a collapsed wall in the boiler room. There was minor damage to the refinery's pipelines and fuel tanks. The closure was part of a precautionary measure. It will be returned to operation after all damage is fixed and final inspection completed.

The Aconcagua refinery west of Santiago was closed for a couple of days for a safety inspection then reopened.

Hospitals and Schools

Within the earthquake-impact zone, 80 hospitals sustained various level of damage. There were no collapsed hospitals.

The earthquake and the tsunami affected about 4000 schools. There were partial collapses, but no completely collapsed schools were reported.

Emergency Response, Recovery, and Social Impact

The military was deployed to restore order in the Concepción area due to violence and looting. It also provided temporary buildings for tsunami victims. About 430,000 homes were impacted by the earthquake and the tsunami.

Damage to roads and bridges delayed delivery of relief materials the first 3 days after the earthquake.

1.3 Organization

This report discusses the following lifeline systems: electric power, communication, water, highways, and emergency response. Lifeline systems within hospitals are also discussed. Most lifeline chapters start with an overview of the system performance, followed by sections that describe the system, its damage, emergency response, and recovery. Each chapter finishes with a conclusions and recommendations.

2
SEISMOLOGY AND GEOTECHNICAL CONSIDERATIONS

A magnitude 8.8 earthquake occurred on an active subduction zone where the Nazca Plate is being subducted under the South American plate. This area has a long history of large magnitude earthquakes and tsunami; as a result, Chile is well recognized for its high seismic hazards. The February 27, 2010 earthquake triggered strong ground shaking over an extensive area that resulted in regional surface deformation, uplift and subsidence, and a near-field tsunami. The first section of this chapter describes the plate tectonic setting, historic seismicity, the expected ground motions from probabilistic maps, seismological characteristics of the earthquake, areas of regional surface deformation, uplift and subsidence, the tsunami inundation maps, tsunami characteristics and impacts, and local geologic conditions.

The second section of this chapter discusses the geotechnical and engineering aspects of the earthquake and how they relate to lifelines. The areas nearest the rupture plane generally experienced a moderate level of ground deformation, ranging from extensive to none in areas by the rivers, coast, floodplains, lowlands, and dunes. The most extensive lifeline damage was observed closest to the coast, with a concentration in areas of poor soils that failed through liquefaction, lateral spreading, or settlement or that were inundated by the tsunami. The level of ground shaking was controlled by the rupture pattern, the proximity to the earthquake rupture, and the local soil conditions. Ground shaking and geotechnical-related ground failures impacted all lifelines and resulted in damage to buildings, bridges, highways, railways, ports, and systems involving water, wastewater, gas and liquid fuel, electricity, and telecommunication as well as contributing to lifeline interdependency issues. Geotechnical-related failures involved soft soil site amplification, bearing capacity and fill failures, slope and dam failures, and liquefaction and lateral spreading failures—the last two occurring especially near waterways. Areas with engineered soils, including industrial facilities on wetland sites, engineered retaining walls, waterfront structures, and ground improvement, appeared to perform well. See the sections below as well as the specific lifeline chapters for more detailed information.

2.1 Geologic Aspects

Plate Tectonic Setting

The Maule, Chile earthquake, which occurred at a converging plate boundary, was a subduction zone earthquake. The earthquake occurred in a region of known high seismicity at the interface between the Nazca plate and the South American tectonic plates. The Nazca plate is converging eastward at a rate of about 65 to 80 mm per year. Near the Maule earthquake, the convergence rate is 80 mm per year (Fig. 2.1).

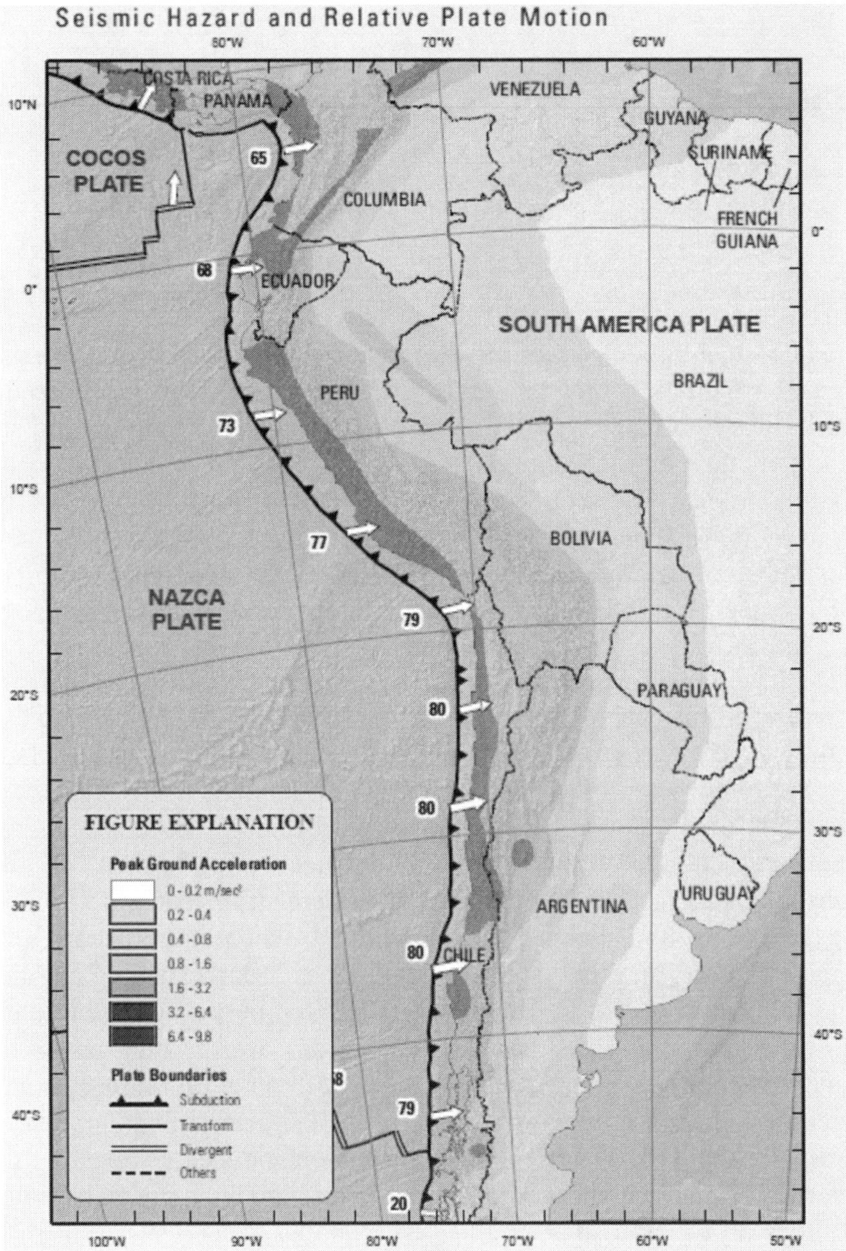

Figure 2.1. Plate Tectonic Setting. (Source: USGS, 2010, http://earthquake.usgs.gov/earthquakes/eqarchives/poster/2010/20100227.php)

Chile has a long history of large earthquakes, including great subduction zone earthquakes of magnitude 8 or larger. Table 2.1 includes all known magnitude 8 and larger earthquakes dating from 1575. Figure 2.2 shows the epicenter of the Maule earthquake and magnitude 7 and larger earthquakes dating from 1900.

Table 2.1. Great Historic Earthquakes in Chile

1575 Valdivia earthquake	8.5
1730 Valparaiso earthquake	8.7
1751 Concepción earthquake	8.5
1835 Concepción earthquake	8.5
1868 Arica earthquake	9.0
1906 Valparaíso earthquake	8.2
1922 Vallenar earthquake	8.5
1943 Coquimbo earthquake	8.2
1960 Valdivia earthquake	9.5
1965 La Ligua	9.0
1985 Santiago earthquake	8.0
1995 Antofagasta earthquake	8.0

Source: Wikipedia, http://en.wikipedia.org/wiki/List_of_earthquakes_in_Chile

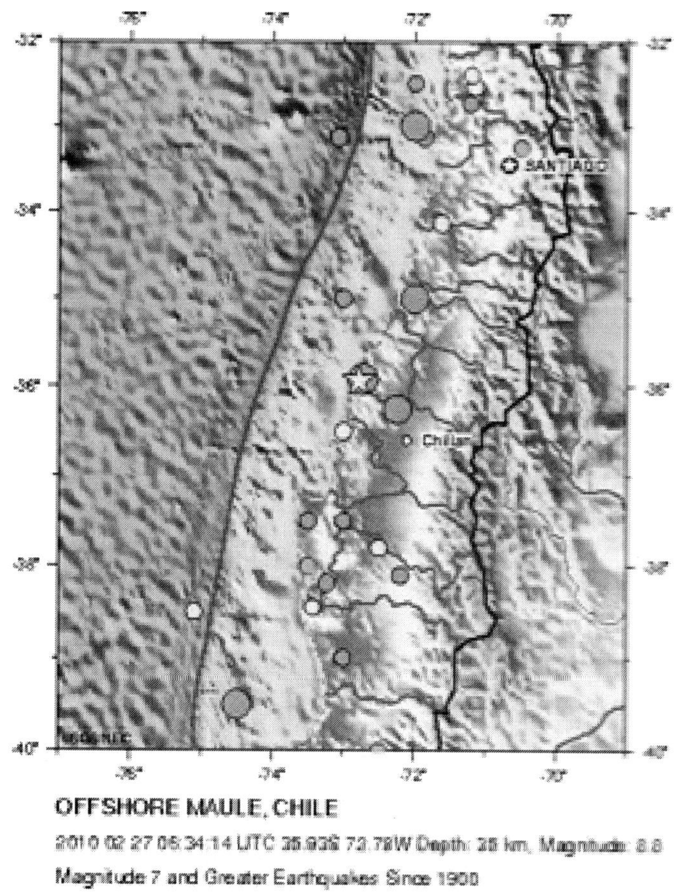

Figure 2.2. Maule earthquake and magnitude 7 and larger earthquakes (Source: USGS, 2010, http://neic.usgs.gov/neis/eq_depot/2010/eq_100227_tfan/neic_tfan_h.html)

Seismic Hazard and Seismicity

Probabilistic ground motion maps have been developed on the basis of known faults, historical seismicity, and likely earthquakes. Chile is considered to have an active plate tectonic setting: It has many mapped active faults, experiences frequent earthquakes, and consequently, has a high seismic hazard. Due to the subduction zone, the hazard is greatest along coastal Chile where high ground motions are expected. The expected ground motion levels reduce going eastward away from the subduction zone (Fig. 2.3). The expected peak ground accelerations with a 10 percent probability of exceedence in 50 years for rock sites vary from more than 4.8 to 2.8 m/s^2 along the coast to 4.0 to 2.4 m/s^2 in the valley (e.g., Santiago and Chillán).

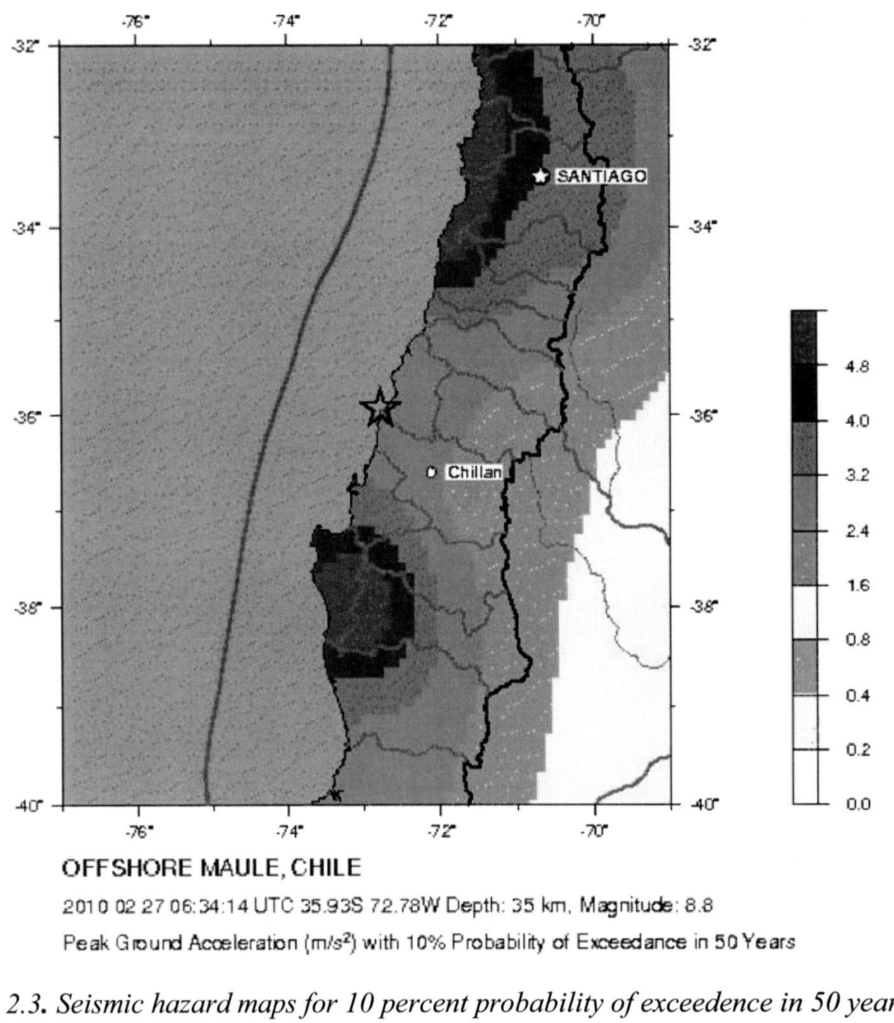

Figure 2.3. Seismic hazard maps for 10 percent probability of exceedence in 50 years (Source: USGS, 2007, http://neic.usgs.gov/neis/eq_depot/2010/eq_100227_tfan/neic_tfan_w.html)

Maule Earthquake Characteristics

On February 27, 2010 at 3:34 a.m. local time (06:36 UTC), there was a moment magnitude (Mw) 8.8 earthquake off the west coast of the Maule Region of Chile (http://earthquake.usgs.gov/earthquakes/eqinthenews/2010/us2010tfan). This earthquake was located offshore at 35.909°S, 72.733°W with a depth of 35 km and had a plate rupture of about 550 km by 150 km. It shook a large region with an estimated 93 percent of Chile's population. The epicenter is located 335 km SW of the capitol city of Santiago and 105 km NNE of the coastal city of Concepción.

Numerous aftershocks occurred during the following months, including 15 Mw 6 or larger aftershocks and more than 50 greater than Mw 5 earthquakes within the first three weeks following the earthquake. Figure 2.4 shows the Maule earthquake epicenter and aftershocks (white circles); past subduction zone, crustal, and intraplate earthquakes; and active volcanoes.

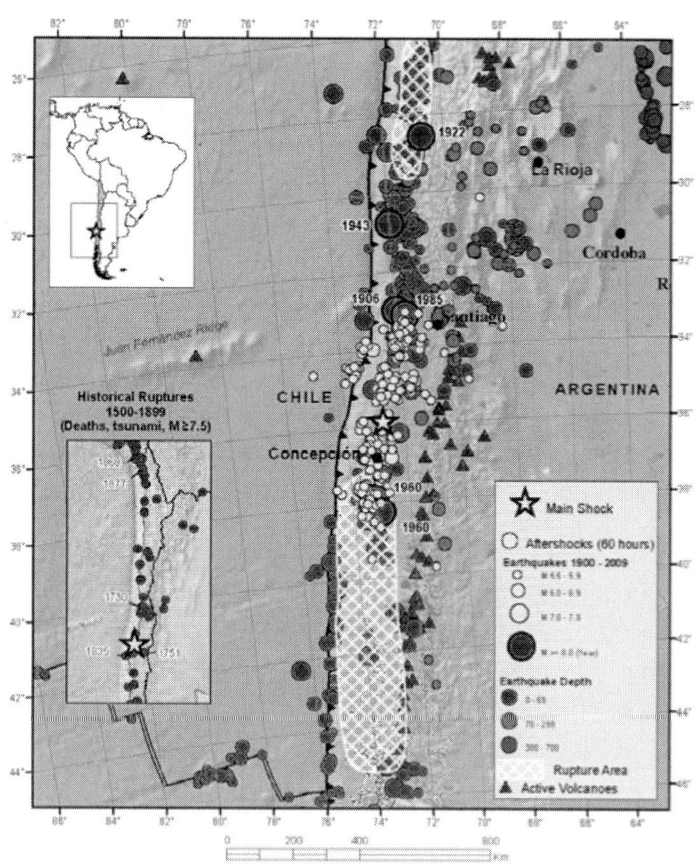

Figure 2.4. Earthquake location map, including main shock, aftershocks, and past earthquakes (Source: USGS Web site, http://earthquake.usgs.gov/earthquakes/eqinthenews/2010/us2010tfan/20100227_Main.pdf)

The Maule earthquake is the fifth largest earthquake in the entire world's recorded history. It occurred immediately north of the largest earthquake recorded to date, the 1960 magnitude 9.5 earthquake, whose northern terminus was south of the Arauca Peninsula. The Maule earthquake ruptured to its north, near the historic 1835 event, which evolutionist Charles Darwin experienced. The Maule earthquake's rupture zone falls between the 1960 event to the south and the 1906 event to the north offshore Santiago.

An estimated 10 m or more of slip deficit had accrued on this segment of the plate boundary since the 1835 earthquake (Ruegg et al. 2009). The USGS finite fault model is consistent with this estimate, showing about 450 cm of displacement at the seafloor surface and approaching 900 cm near the hypocenter (USGS Web site, UNAVCO Web site). The surface projection of the fault plane slip distribution is shown in Figure 2.5. During this earthquake the coast of Chile sprang westward towards the ocean, in some places by as much as 6 m (19 feet) (CIT Web site).

The Maule earthquake released a large amount of stored seismic strain over a matter of minutes, which led to permanent land level changes. Figure 2.6 presents a map of the surface deformation model, which is based on the fault plane slip model (CIT Web site). Coastal regions experienced both uplift and subsidence (Fig. 2.7). Land-level changes ranging between 2.5 m and 1 m were observed in vertically displaced coastal and river markers along approximately 500 km of the rupture length (Fraias et al. 2010). A north-south hinge line located 120 km eastward from the trench separates uplifted areas to the west from subsided regions to the east (Fraias et al. 2010).

Ground Shaking

The earthquake resulted in maximum recorded peak ground accelerations of 0.65 g in the Concepción area, which experienced a bracketed duration of more than 60 seconds of strong shaking of engineering significance. The USGS shake map shown in Figure 2.8 illustrates the approximate fault rupture area and estimated ground motions described in terms of peak ground accelerations, peak ground velocities, and Modified Mercalli Intensities (MMI).

The maximum ground shaking intensity was estimated at MMI VIII and was experienced in a number of coastal and inland communities. The new USGS PAGER (Prompt Assessment of Global Earthquakes for Response) product provides rapid estimates, including population exposure overlain on an MMI map. It provides past earthquake information as well as order of magnitude fatalities and economic loss estimates from the current earthquake with associated probabilities. Finally, it provides an alert level at the top. It does not include estimates of secondary effects, such as those from tsunamis, landslides, and liquefaction.

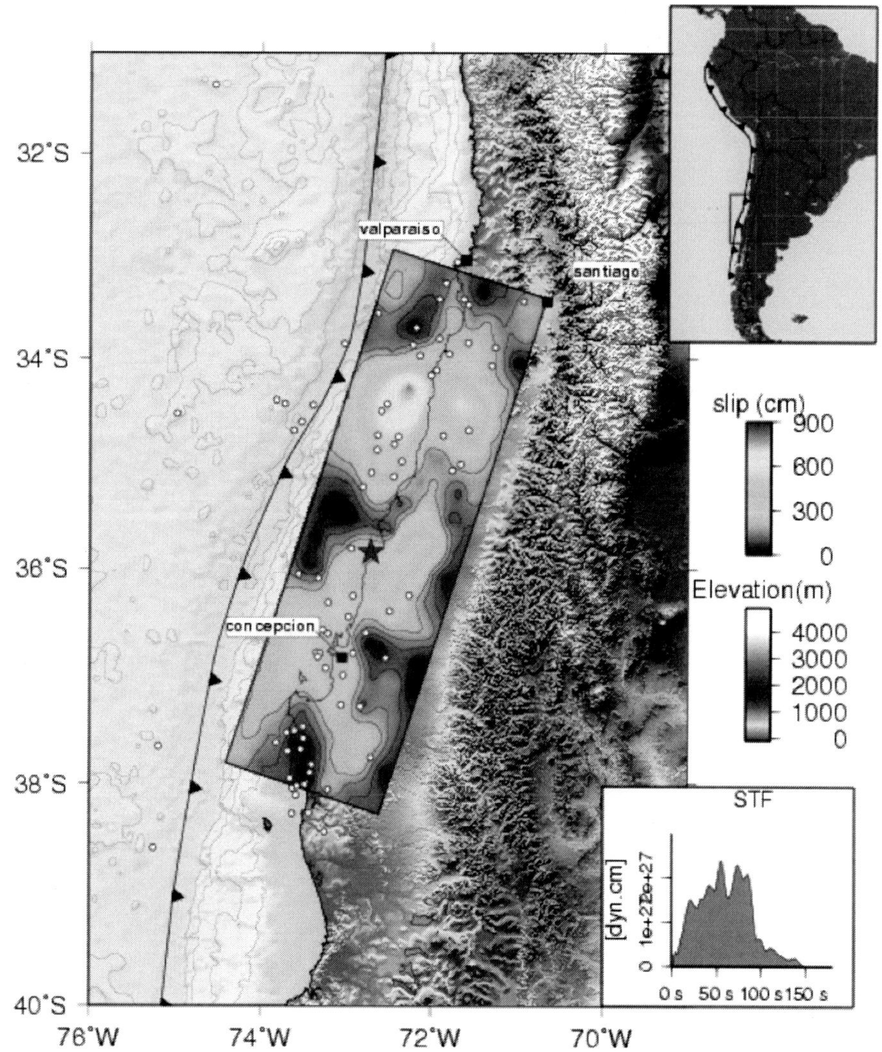

Figure 2.5. Map view of slip distribution with the epicenter (red star) and aftershocks located by USGS during the day following the earthquake (white dots) (Source: CalTech Tectonics Observatory, 2010)

For the Maule earthquake, PAGER indicates that more than 15 million people were exposed to areas of MMI IV to MMI VIII. PAGER results estimated that there was a 40 percent probability that 100 to 1000 fatalities would be caused by earthquake shaking and that there was a 45 percent probability of economic losses between $10 billion and $100 billion, and it issued a red alert level (Fig. 2.9). PAGER information is invaluable for disaster responders and others who need quick situational assessments.

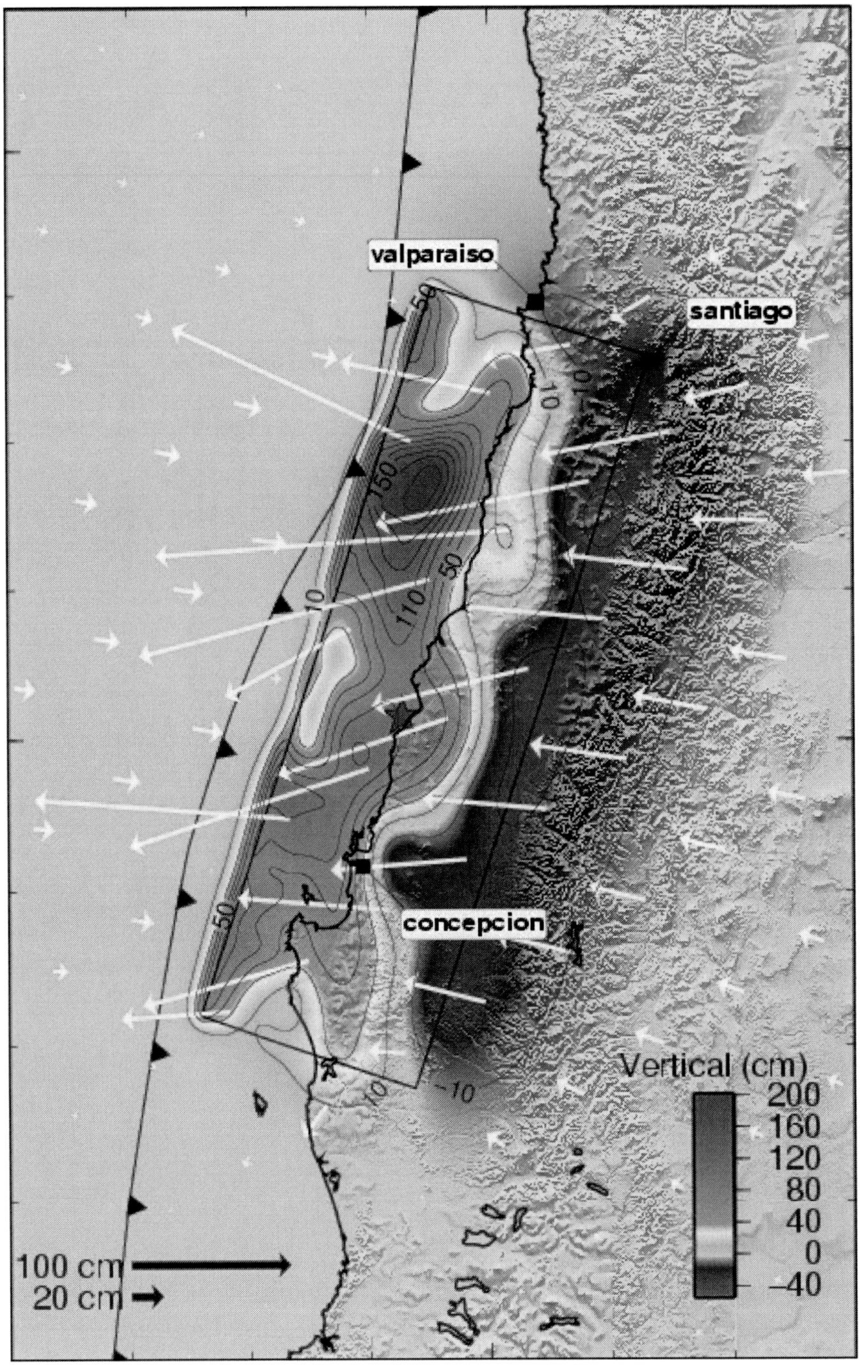

Figure 2.6. Surface deformation predicted from the slip model, shown in Figure 2.5. The vertical component of displacement is given by the color scale and the horizontal motion by the arrows. (Source: CalTech Tectonics Observatory, 2010)

Figure 2.7. Coastal areas experienced permanent uplift and subsidence. This pier was uplifted, which impacted pier operations.

Geological Setting

The geologic conditions vary in the large epicentral region (Fig. 2.6 and 2.7) from mountainous terrain and valleys to river and coastal terrains. The coastline varies from low mountains and marine terraces to coastal dunes and river mouths with loose fluvial soils and shallow groundwater conditions. Deltaic lowlands form at the river mouths by fluvial deposition. Similarly, coastal lowlands form along the shoreline from aggrading shallow marine deposits. Sand dunes form along ancient and current shorelines and rivers from wind deposits and often overlie sandy lowlands. Lowlands, composed of mostly soft cohesive, organic-rich soils, form between dunes and river environments. The Bio Bio River flows through Concepción to the Pacific Ocean. The flood plains consist of layers of loose sandy and soft cohesive soils.

Certain localized areas were subjected to ground motion amplification due to softer soils and/or topography as well as focusing of high seismic energy. The groundwater conditions were favorably low because the earthquake occurred towards the end of summer. As a result, fewer landslides and ground related failures related to liquefaction, lateral spreading, and bearing capacity occurred.

Crustal Faults

Numerous inland crustal faults have been mapped in western Chile. A moment magnitude 6.9 occurred on a crustal fault on March 11, 2010, 40 km southwest of Pichilemu, which is about 150 km southwest of Santiago along the coast (Fig. 2.10). This earthquake is not considered an aftershock of the Maule earthquake because it did not occur on the subduction zone and exhibited normal faulting motions. However, it was caused by the regional stress conditions. The damage was most severe in Pichilemu, but many houses in the nearby towns of Espinillo and Rodeíllo were also destroyed. There were two reported fatalities (http://en.wikipedia.org/wiki/2010_Pichilemu_earthquake).

Chilean Tsunami Hazards

Chile has a long history of tsunamis associated with the active subduction zone and resulting earthquakes; thus, it has a high tsunami hazard along its entire coastline. Tsunami hazard maps have been produced for much of Chile's coastline by SHOA and are available to the public (Fig. 2.11).

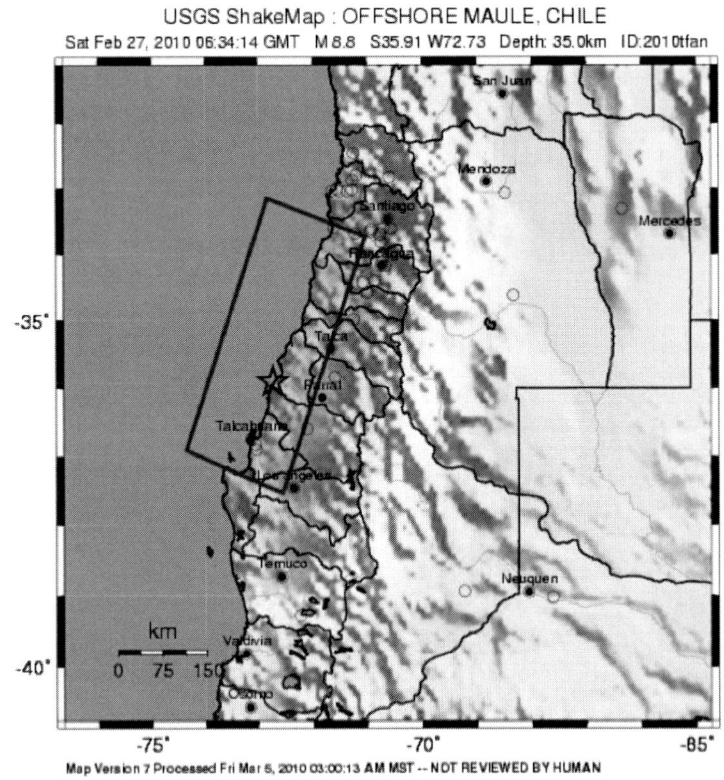

PERCEIVED SHAKING	Not felt	Weak	Light	Moderate	Strong	Very strong	Severe	Violent	Extreme
POTENTIAL DAMAGE	none	none	none	Very light	Light	Moderate	Moderate/Heavy	Heavy	Very Heavy
PEAK ACC.(%g)	<.17	.17-1.4	1.4-3.9	3.9-9.2	9.2-18	18-34	34-65	65-124	>124
PEAK VEL.(cm/s)	<0.1	0.1-1.1	1.1-3.4	3.4-8.1	8.1-16	16-31	31-60	60-116	>116
INSTRUMENTAL INTENSITY	I	II-III	IV	V	VI	VII	VIII	IX	X+

Figure 2.8. The USGS shake map shows the approximate fault rupture area and estimated ground motions described as peak ground accelerations, peak ground velocities, and Modified Mercalli Intensities (Source: USGS, 2010, http://earthquake.usgs.gov/earthquakes/shakemap/global/shake/2010tfan/)

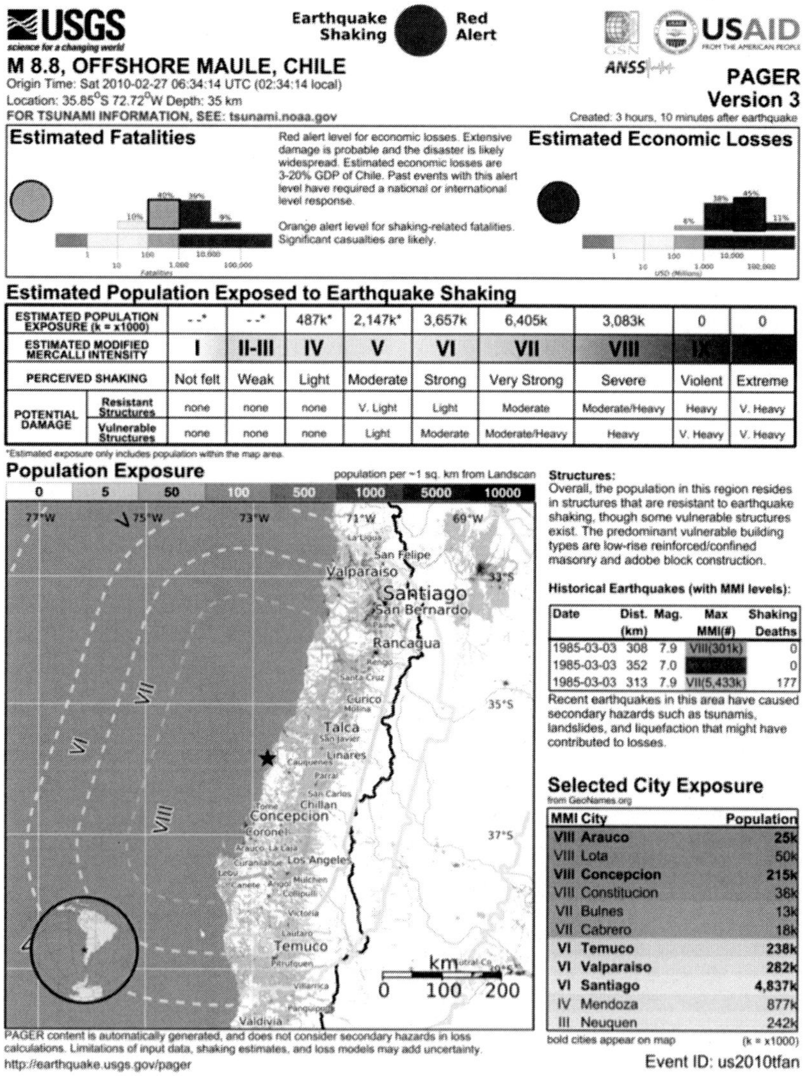

Figure 2.9. The USGS PAGER product shows MMIs, population exposure, estimated fatalities, economic loss, and a red alert level. (Source: USGS, 2010, http://earthquake.usgs.gov/earthquakes/pager/events/us/2010tfan/exposure.png)

One of the most notable tsunamis in Chilean history was triggered by the 1751 Concepción earthquake. This resulted in the relocation of the city 14 years later from its original site near the present day town of Penco to its present location. The current town of Penco was damaged by the tsunami triggered by this Maule earthquake.

The 1751 earthquake and tsunami was one of the strongest and most destructive recorded in Chile's history. The records from personal accounts indicate that the earthquake lasted about 6 minutes and was so intense that the residents could not remain standing. The earthquake and subsequent tsunami reportedly destroyed all of the buildings in coastal Concepción, which at the time was still recovering from a 1730 earthquake and tsunami. The earthquake also struck the central valley, destroying the cities of Chillán, Cauquenes, Curicó, and Talca. One of the most affected cities near Concepción was Chillán, where the entire city moved about 15 blocks and was destroyed, and the river changed its course. Although no other major damage was reported in Santiago, the tower of the cathedral was destroyed by the tremor.

The 1751 tsunami arrived as a series of waves some 10 to 40 minutes after the earthquake. The seawater receded more than 1 km before three to five wave sets inundated the low-lying coastal land. The height and force of each wave increased, with the last was the most disastrous. The 1751 tsunami destroyed the new settlement at the Juan Fernández Islands, where 35 people died (Wikipedia Web site, http://en.wikipedia.org/wiki/1751_Concepci percentC3 percentB3n_earthquake).

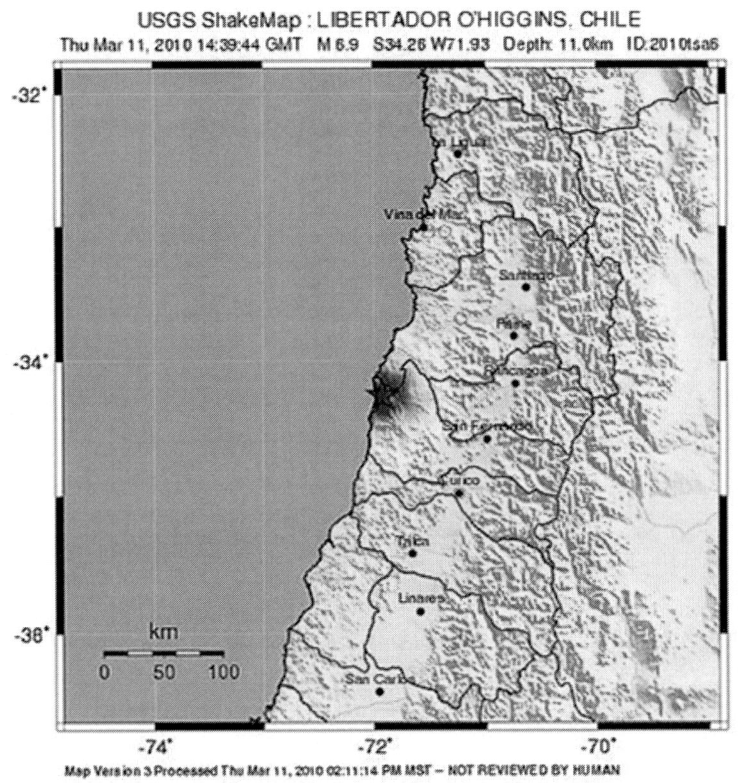

PERCEIVED SHAKING	Not felt	Weak	Light	Moderate	Strong	Very strong	Severe	Violent	Extreme
POTENTIAL DAMAGE	none	none	none	Very light	Light	Moderate	Moderate/Heavy	Heavy	Very Heavy
PEAK ACC.(%g)	<.17	.17-1.4	1.4-3.9	3.9-9.2	9.2-18	18-34	34-65	65-124	>124
PEAK VEL.(cm/s)	<0.1	0.1-1.1	1.1-3.4	3.4-8.1	8.1-16	16-31	31-60	60-116	>116
INSTRUMENTAL INTENSITY	I	II-III	IV	V	VI	VII	VIII	IX	X+

Figure 2.10. The USGS shake map shows the magnitude 6.9 Pichilemu earthquake on a crustal fault, estimated ground motions described as peak ground accelerations, peak ground velocities, and Modified Mercalli Intensities (Source: USGS, 2010, http://earthquake.usgs.gov/earthquakes/shakemap/global/shake/2010tsa6/download/tvmap.jpg)

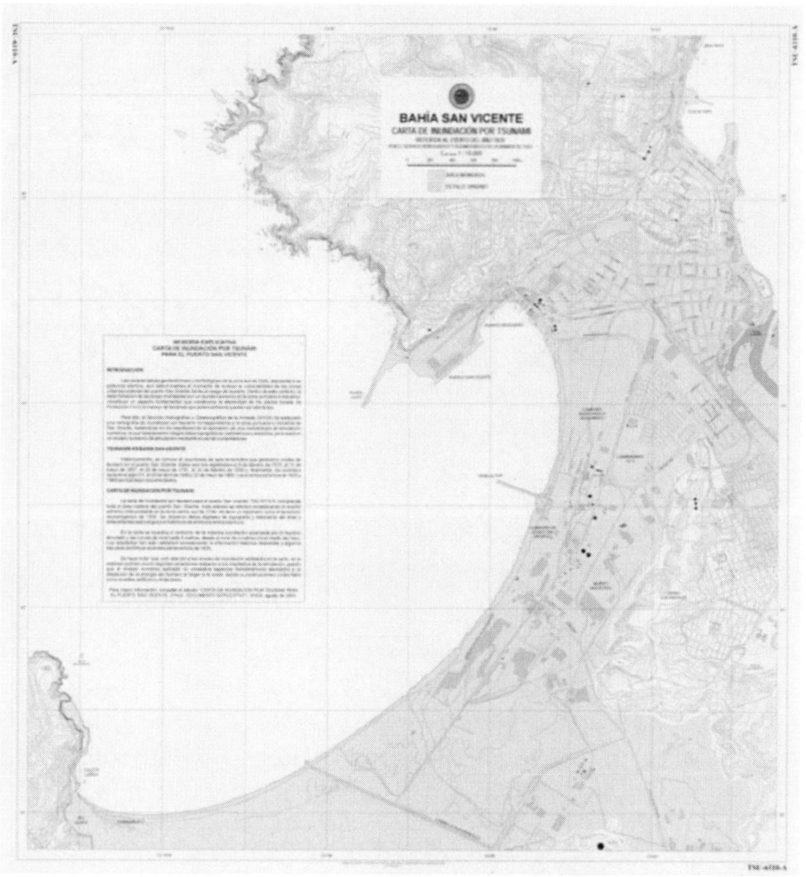

Figure 2.11. Example of a SHOA-produced tsunami hazard map of a San Vicente port, Bahia San Vicente (Source: SHOA, 2004, http://www.shoa.cl/servicios/citsu/citsu_ing.php)

Maule Earthquake's Tsunami Characteristics

A tsunami was generated by the fault rupture of the seafloor and propagated in all directions from the rupture. Figure 2.12 shows NOAA's modeling results for the expected tsunami wave amplitude and travel time for the Pacific Ocean. The tsunami waves hit the low-lying Chilean coastline as well as distant shores across the Pacific Ocean. The tsunami resulted in 200-some fatalities in Chile's near field zone but was harmless in the rest of the Pacific region.

The tsunami had various characteristics up and down Chile's coast. The variations in wave characteristics were primarily a function of the fault rupture, bathymetry, and coastal topography. Tsunami waves were measured on coastal tide gages to the north, south, and west of the epicenter. The tsunami height was approximately 1 m at locations from latitudes 30 to 45 degrees, and greater than 4 m at locations from about latitudes 32 to 40 degrees, which is where the most severe damage occurred.

Figure 2.12. NOAA tsunami model showing amplitude and travel time (Source: NOAA, 2010, http://nctr.pmel.noaa.gov/chile20100227/)

The tsunami was reported to have initiated during a low tide, and subsequent incoming waves occurred during an incoming tide. Three main surges were reported along the south coast. Maximum run-up heights of approximately 20 m along cliff faces and a maximum inland distance in drainages of approximately 12 km from the shoreline were reported by Japanese scientists performing tsunami reconnaissance. At the time of this writing, no published reports on tsunami measurements were found.

The TCLEE team observed tsunami inundation effects on lifelines in many coastal areas. Damage to specific lifelines is described in more detail in the subsequent sections of this report. The amount of damage ranged from minor to severe throughout the affected area and included destroyed lifeline infrastructure—bridges and pavement, electrical and communication systems, water and wastewater systems, buildings, ports, and agriculture. The damage was primarily a function of the tsunami inundation height, the tsunami flow velocity, strength of the structure, and the ability for structures to withstand tsunami forces and impact from floating debris. Figure 2.13 shows a multiple span bridge in Constitucion that was lifted by the tsunami and translated about 1 ft. inland. The bridge girder on the north end temporarily floated, and about six bents were moved. During the TCLEE team visit on April 16, 2010, the bridge traffic was being controlled to one lane of traffic. Team members witnessed official search activities at the north abutment for missing persons (see "Bridge" section for more information).

In general, wood and other light frame buildings suffered extensive tsunami damage. Reinforced concrete buildings are inherently more resistant to tsunami loads, and better performance was observed in these structures.

The 2004 Sumatra earthquake/Indian Ocean tsunami helped to increase awareness and preparedness of tsunamis worldwide, including this one in Chile. Immediately after the earthquake, tsunami warnings were issued by NOAA as well as Chilean organizations. Some of Chile's tsunami warnings were broadcasted by radio. One of the radio warnings issued by the country's president prematurely announced that the tsunami threat had passed when, in fact, it had not.

Tsunami notifications across the Pacific Ocean resulted in thousands of people evacuating from low-lying coastal areas in Hawaii, other Pacific Islands, and other Pacific coasts. NOAA's predicted arrival tsunami times were accurate; however, the wave amplitude predictions have much more uncertainty. In some places, water receded ahead of the tsunami, but the wave amplitudes were only centimeters to 1 m high and did not cause any damage. In contrast, the 1960 Chilean earthquake generated tsunami waves of up to 10 m throughout the Pacific Ocean, devastating parts of Hilo, Hawaii, and causing damage in New Zealand.

Figure 2.13. Tsunami damage to a bridge in Constitucion.

2.2 Geotechnical Engineering Aspects

Ground Shaking

The ground shaking that occurred during this earthquake was highly variable over the area affected by the rupture zone. The quake lasted approximately 3 minutes and strong shaking (>0.05 g) was felt in most areas for approximately 60 to 90 seconds. According to records collected by the University of Chile, maximum peak horizontal accelerations varied between 0.14g and 0.17g in Valdivia and Santiago, respectively, to 0.58g to 0.65g at sites in Concepción. Figure 2.14 shows the recorded motions at Constitucion, a coastal town north of Concepción, which also experienced strong ground shaking. The record shows a peak ground acceleration of approximately 0.5 g. Recorded vertical accelerations throughout the earthquake zone ranged between 0.05 g and 0.60 g. More detailed information about ground shaking can be found in reports produced by the University of Chile (www.terremotosuchile.cl/) and the GEER (www.geerassociation.com) and PEER (http://peer.berkeley.com) reconnaissance teams.

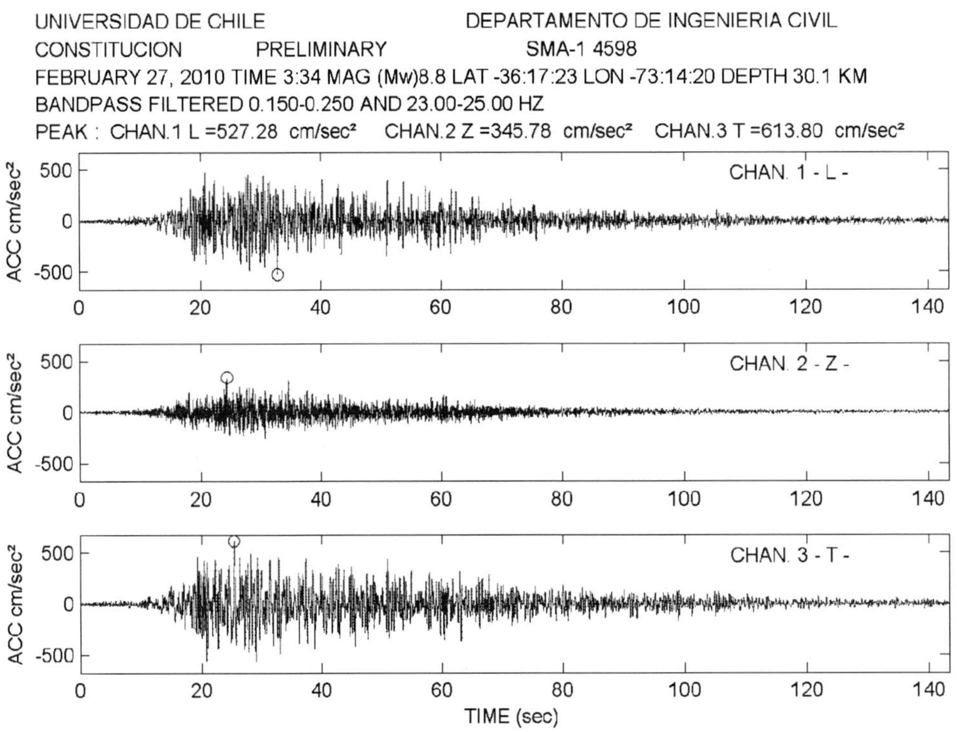

Figure 2.14. Accelerogram from site in coastal Constitucion showing strong shaking for more than a minute. (Source: Boroschek, R., Soto, P. Leon, R. Records from Maule Earthquake Mw = 8.8 February 27 2010. University of Chile Renadic, 10/05 Rev. 2 October, 2010.)

In addition to strength of shaking, the type of shaking, ground motion period, and duration of the shaking also had significant effects on the resulting damage and soil behavior during the earthquake. Typical to other observed subduction zone earthquakes, the large rupture zone resulted in strong shaking that lasted for more than 60 seconds in many places, and the quake itself lasted for more than 2 minutes. This resulted in a high number of significant cycles and, thus, more severe damage. Fatigue failures in structures, higher pore water pressure generation in soils (which leads to more liquefaction and lateral spreading), and a higher potential for strength loss in non-liquefiable soils are common in these types of earthquakes and were observed during the Maule earthquake. The ground motion response spectra are also different from those observed during shallower crustal earthquakes. The response spectra are shifted toward longer periods. This is illustrated in Chapter 6 in the discussion relating ground motions to the damage to electrical structures. Examples of response spectra are also included in the discussion in this section and in the University of Chile, GEER, and PEER reports.

Shaking was also affected by subsurface conditions and topographic effects. These effects resulted in highly variable damage to bridges, highways, and other lifelines in the affected area. Based on the microzonation maps, site soil amplification occurred in the industrial area of Ciudad Empresarial in Santiago. Basin effects were also likely a factor in this area. Topographic and subsurface conditions resulted in larger ground accelerations, which in turn resulted in greater structural and architectural damage to the recently constructed buildings in the area compared to structures in surrounding areas. Some of these effects are illustrated in Figures 2.15a and 2.185b.

Figures 2.15a and 2.15b. Structural damage in Cuidad Empresarial in Santiago. Soil amplification resulted in increased damages in this area of Santiago.

In the city of Concepción, ground motions recorded on bedrock appear rich in low frequency energy and had uncorrected maximum recorded peak ground accelerations of 0.65 g in the north-south direction, 0.58 g in the east-west direction, and 0.60 g in the vertical direction. We anticipate that peak accelerations on alluvial sites are significantly lower due to deamplification from deeper soft and liquefiable soils; however, no recordings have been obtained. Examples are Concepción's wastewater plant and the Empresa Nacional del Petróleo (ENAP) refinery, which were developed on sites adjacent to the Bio Bio River. Damage due to soil strength loss, liquefaction, and lateral spreading were observed; however, structural and architectural damage were less than expected based on peak ground accelerations recorded in other areas of Concepción. An additional source of ground shaking was the 10 ft. of coseismic westward movement (rebound effects) that occurred along the plate boundary. These effects are currently being researched.

Liquefaction and Lateral Spread

During earthquake shaking, loose, saturated sandy soils can contract, which can lead to an increase of pore water pressure. When pore water pressure increases, the effective stress reduces, and soils can "liquefy." Liquefaction results in loss of soil strength and, thus, foundation bearing capacity failures, slope stability issues, widespread settlements, and other damage. Sand boils like that shown in Figure 2.16 are a common indication of liquefaction. Liquefaction is often associated with lateral spreading and permanent ground displacements in the down gradient direction. Because it occurs in loosely deposited saturated soils, liquefaction and lateral spreading most often happen in lowlands with high groundwater conditions and near waterways. Several coastal and river vicinities experienced extensive liquefaction, resulting in sand boils, fissures, and lateral spreading. Lateral spreading in port areas occurred where long shallow sloping coastlines were affected by lateral spreading. In several areas the lateral spreading continued back 100 to 200 ft. from the water line. An example of lateral spread damage is shown in Figure 2.17. Bridges, highways, and lifelines at water crossings were damaged due to liquefaction and lateral spreading. In Concepción alone, all five of the major river crossings experienced damage due to liquefaction and lateral spread. Because of this damage, traffic was severely restricted in the Concepción area at the time of the TCLEE reconnaissance and riots protesting these restrictions were occurring.

The earthquake occurred at the end of summer, during the driest period of the year. Had the earthquake occurred during the winter or a period with higher groundwater levels, the shaking and the resulting liquefaction and lateral spreading would have caused more damage.

Specific cases of lateral spreading and liquefaction are detailed in the lifeline sections.

Figure 2.16. A sand boil near the abutment of the failed Juan Pablo II Bridge over the Bio Bio River in Concepción.

Figure 2.17. Liquefaction and lateral spread damage to the park area and sheet pile wall near the Port of Coronel.

Slope Failure

Many slope failures occurred during this earthquake. However, the slope failures were localized and infrequent given the magnitude and duration of the earthquake and the geography of the area affected by strong shaking. Areas that experienced earthquake-triggered landslides were mostly areas with weak slopes and previous landslide terrain. Road cuts in both soils and rock and steep slopes that were marginally stable before the earthquake were commonly destabilized due to the ground shaking. Frequent small slides were observed in road cuts throughout the area the TCLEE team explored, such as that shown in Figure 2.18. Further, slope instabilities were also observed where slopes had been undercut by river or wave action. The National Service of Geology and Mineralogy (Sernageomin) documented many landslides in residential areas that had minimal effects on any major lifelines.

Chile has a history of frequent landslides in coastal regions and areas where steep slopes are affected by heavy rains. This earthquake occurred at the end of a very dry summer. Many existing slides that had a previous history of instability were reactivated due to the ground shaking. One example observed by the TCLEE team is shown in Figure 2.19. This slide east of the town of Iloca is in a zone of meta-stable landslides. Due to post-earthquake rebuilding nearby, this slide is now being monitored by Sernageomin in anticipation of the wet season.

Figure 2.18. Slope instability and rockslide in a residential area of Talcahuano taken days after the earthquake by Sernageomin staff.

The low frequency of slope instability in the coastal areas as well as areas farther inland is attributed to the season in which the earthquake occurred. Had the earthquake occurred during the wet season, significantly more damage from slope instabilities would have been observed.

Tsunami Damage

In areas where the tsunami caused most of the damage, it was difficult to determine the direct cause of damage to structures and lifelines. It is likely that shaking damage was present and that liquefaction and lateral spreading occurred in these areas; however, most evidence was destroyed by the waves and inundation. Two phenomenona observed in the tsunami areas are worth mentioning from a geotechnical standpoint. The first observation was that storm and sanitary sewer lines were uncovered and lifted from their original locations. This was observed in several areas where the tsunami was forceful or large inundations occurred. Most likely the combination of the upward buoyant forces and the scouring due to wave action caused the lines to be uncovered and allowed them to move vertically as well as laterally as shown in Figure 2.20.

The second phenomenon was observed south of the coastal town of Iloca. In this area large sections of asphalt pavement were lifted by the uplift forces caused by tsunami waves. They floated off roadway embankments and were deposited by the water above and below the original embankments. The asphalt concrete acted as a large crust and was folded and deposited in layers as shown in Figure 2.21.

Figure 2.19. Reactivated landslide above the town of Iloca. Temporary housing can be seen built directly under the visible landslide topography.

Figure 2.20. Storm sewer damaged by tsunami forces south of the town of Iloca.

Figure 2.21. Asphalt that was lifted and transported by the tsunami waves south of the town of Iloca.

Engineering Aspects

Subsurface Conditions in Earthquake Codes and Ground Improvement in Construction

Chilean building codes have considered seismic design since the 1930s. The general soil characteristics and geology have been mapped for the larger cities including Concepción and Santiago. These soil conditions have been separated into four seismic zones to account for soil amplification effects and recently have been incorporated into the national building codes so that site subsurface soil conditions are considered during design of buildings. Maps of the soil conditions are included in the Chilean National Building Code for design purposes. No official codes govern lifeline or bridge design and construction; however, it is evident that some consideration for seismic design has been incorporated into the design of certain lifeline structures based on owner policies, local experience, and seismic history.

Because site-specific seismic considerations have been included in the design of buildings structures in Chile, ground improvement techniques are widely used to improve soft and liquefiable soils. Based on our conversations with government officials and private lifeline owners, the majority of ground improvement consisted of the removal of the soft or loose material and replacement with engineered fill. These fills were often tens of feet thick. Vibratory compaction is used in the placement of these fills, as shown in Figure 2.22. We understand this method was used in wetland sites near Concepción, ports, earthen dams, levees, bridge abutments, road embankments, and cut-and-fill roads. For the most part, engineered fill soils performed well.

In Concepción, geotechnical ground improvement was conducted in two limited areas at the ENAP oil refinery. Stone columns and micropiles supporting heavy equipment were reported to perform well with no observed ground movement or settlement in the vicinity.

Information provided by government officials and other Chilean professionals indicated that deep dynamic compaction had been used in areas of Concepción and soil mixing had been utilized for a new high-rise structure in Valparaiso. It was reported that both systems performed well. TCLEE was unable to confirm the location of these areas; however, the large wrecking ball shown in Figure 2.23 was rumored to have been used for deep dynamic compaction for a riverside housing development in Concepción.

Retaining Walls

The main failure mechanisms for retaining walls during this earthquake were bearing capacity and lateral spreading failures that occurred in soft or liquefiable bearing soils under the retaining walls. Many of the ports experienced failures of minor block walls due to soil movement beneath the walls as shown in Figure 2.24. Liquefaction and lateral spreading also affected soldier pile and sheet pile walls at ports that did not extend past liquefiable and soft layers. An example is shown in Figure 2.25. No structural failures of retaining walls were observed by the TCLEE team nor were any reported by any of our reconnaissance partners.

Figure 2.22. Compaction operations for a large new fill for a new port area at the Port of Lirquen

Figure 2.23. The large wrecking ball rumored to have been used to perform dynamic compaction for the public housing along the Bio Bio River in Concepción

Figure 2.24. A block wall damage due to subgrade failure at the Port of San Vincente

Figure 2.25. Shallow sheet pile wall failure due to poor soils at the La Poza port (Courtesy of the Ministry of Public Works, Ports Department)

In general, retaining walls supporting bridge approaches and abutments performed well if founded in competent soils. A large rock anchor/soil nail wall was observed above the Ciudad Empresarial in Santiago. No visible damage was noted during our reconnaissance.

Geotechnical Effects on Lifeline Systems

Electrical

The electrical system was affected by ground shaking, amplification, soft soils, and lateral spreading. One of the towers for the main electrical transmission line across the Bio Bio River in the Concepción area collapsed due to liquefaction and bearing capacity failures of the supporting soils. This tower is shown in Figure 2.26. The adjacent towers were pulled down by the failed tower. Further, portions of the distribution systems located along waterways in areas of liquefaction, lateral spreading, and soft soil ground amplification were damaged during the earthquake, such as the substation shown in Figure 2.27. Severe shaking caused buildings and other structures to collapse, which took down power distribution lines. In tsunami zones the power distribution lines were totally demolished. A more detailed discussion of the effects on the electrical system is included in Chapter 6.

Water/Wastewater

The water and wastewater systems were heavily affected by liquefaction and lateral spread because the main treatment plants and structures as well as large portions of the transmission and distribution systems are located in alluvial or coastal areas near waterways and were supported on soft and liquefiable soils.

Figure 2.26a and 2.26b. Power transmission tower collapse due to soft soils (Courtesy of TransElec Power Company)

Figure 2.27. Lateral spread at a substation near the port of San Vincente

In the Concepción area, ground shaking, soft soils, and lateral spreading accounted for 72 breaks in water mains and more than 2000 breaks in the distribution system. Twenty-five out of 50 water pump stations failed. Water intake areas also were affected by lateral spreading because they were located in vulnerable areas close to waterways. Also in Concepción, one of the main wastewater collection systems failed due to liquefaction and lateral spreading and reportedly will require more than 10 months to repair.

The Concepción wastewater treatment plant is located adjacent to the mouth of the Bio Bio River on alluvial soils. This site experienced extensive liquefaction, including sand boils, fissures, settlement, and lateral spreading, which led to extensive damage and will reportedly require months to repair (Fig. 2.28).

Further information can be found in Chapter 8.

Figure 2.28. Lateral spreading and shaking damage at the water treatment plant in Concepción

Ports and Airports

As mentioned in the liquefaction and lateral spreading section above, the ports observed by the TCLEE team were located in areas of very gradually sloping shorelines and many utilized off-shore piers for the unloading and loading of goods. Many of the ports were constructed with short (<10-foot tall) walls that were founded on or in liquefiable and soft soils, such as the San Vincente port (Fig. 2.29). During the earthquake, a large number of these walls were destroyed by lateral spreading or bearing capacity failures. Container areas and storage areas as well as buildings within the lateral spread zone were affected by large-scale movements toward the water. These movements often extended 100 to 200 ft. from the water line. The industrial wastewater treatment plant at the industrial port of Coronel consisted of a partially underground control building that experienced between 1.5 and 3 ft. of uplift due to buoyancy forces in liquefied soils.

Figure 2.29. Lateral spread damage at San Vincente container port (Courtesy of the Ministry of Public Works—Ports Section)

Lateral spreading and liquefaction also damaged the piers that extended out into the water. Those most heavily damaged were the older piers where the foundations did not extend deep enough into competent material (see for example, Fig. 2.30). These piers were often used for the fisheries industry, which was hit hard by pier failures. It will take months to years for the industry to recover the use of all of its ports and piers. The Port of Coronel had recently constructed a partially base isolated pier that performed very well during the quake.

The airports at Santiago, Concepción, and Los Angeles were located in areas where ground amplification caused significant shaking. The majority of the damage at the airports was nonstructural; however, there was a significant loss of serviceability. The terminal at the Santiago airport was not fully operational until 6 weeks after the earthquake. More information on damage to ports and airports is provided in Chapters 4 and 9.

Transportation

Railways, bridges, highways, and roadways within zones of ground amplification, on soft or liquefiable soils, or where lateral spread occurred were severely affected by the earthquake. Large critical structures such as the Llacolen and Juan Pablo II bridges in Concepción and several Ruta 5 bridges as well as hundreds of smaller bridges were affected when abutments and approaches were damaged by ground shaking, lateral movement, and soft soils.

Figure 2.30. The damaged fishing pier at the Port of Coronel

Figure 2.31. Bridge collapse due to rotation on the Amerigo Vespucio around Santiago (Courtesy of the Ministry of Public Works, Bridge Department)

Figure 2.32. Railway tracks at Port of Lirquen snapped due to permanent ground deformation

Strong shaking caused many bridges with insufficient lateral restraints to rotate off their supports. In several cases this resulted in total collapse, including several new structures along Ruta 5 and the highway that circles Santiago (Fig. 2.31). Damages observed varied greatly throughout the earthquake zone due to topographic effects, soil amplification, and other topographic effects. Strong shaking that resulted in ground deformation also affected railways. Railway tracks at the Port of Lirquen snapped in a few places between welds, expansion joints, and mid-section (Fig. 2.32).

Damage along railways and highways were also caused by failures within and beneath embankment fills. In areas of weak and liquefiable soils, bearing capacity failures caused embankments and approach fills to fail (Fig. 2.33). Bridge bents were also affected by bearing capacity failures in poor soils along the spans such as those shown in Figures 2.34 and 2.35. An example of this type of failure was the Juan Pablo II Bridge over the Bio Bio River in Concepción (Fig. 2.34).

Damage was also caused by lateral spreading at approach abutments and culvert crossings near waterways and in areas of alluvial soils. The Llacolen Bridge in Concepción suffered damage when the soil under the eastern abutment moved toward the Bio Bio River (Fig. 2.36). The bridge was reopened to restricted traffic with a temporary Bailey bridge; however, major repairs will be required. Near Talcahuano, the railroad was also damaged by lateral spread.

Further information on transportation is included Chapter 3.

Figure 2.33. Liquefaction and lateral spread damage on an approach embankment along Ruta 5, 130 kilometers northeast of Concepción

Figure 2.34. Bent bearing capacity failure at the Juan Pablo II bridge in Concepción

Figure 2.35. Bent failure due to soil failures in Tubul (Courtesy of the Ministry of Public Works, Bridge Department)

Figure 2.36. Lateral spread damage to the northern approach to the Llacolyn Bridge in Concepción. The arrows indicate the section fitted with a Bailey bridge to reopen.

Telecommunications

The telecommunications industry was most affected by shaking intensity where towers and equipment were damaged due to inadequate anchorage. Systems located in areas of poor soils were also damaged. Information collected indicated that at least one tower collapsed due to bearing capacity failure in poor soils. Further information on telecommunications is included in Chapter 7.

Natural Gas and Liquid Fuel

Natural gas and liquid fuel importing areas, crude oil off-loading areas, refineries, and storage were often located in areas near ports and waterways that were affected heavily by regional and coastal uplift, liquefaction, and lateral spread. In Concepción at the port of San Vincente, the existing ENAP pier, a pier that is currently under construction, and at least one of the two crude intake lines (Fig 2.37) for the main ENAP refinery in the region were damaged. At the location of the two pipelines, the lateral spread resulted in more than 1 m ground movement horizontally towards the water, and the area was also vertically lifted by 0.5 m. Due to public pressures, ENAP replaced these pipelines with ones buried deeper. Water intake facilities and water treatment facilities for refineries and other systems were also damaged due to soft soils and lateral spreading in waterfront areas. Damage to piers and associated on-shore facilities will take several months to repair.

The ENAP refinery in Concepción experienced relatively moderate to strong shaking and reasonably minor damage. Evidence of shaking can be seen by sloshing stains on the storage tanks at and adjacent to the main refinery as seen in Figure 2.38. The main damage within the main refinery occurred during shaking to fin fans, the derrick, and heaters, which were expensive and difficult to replace. The linings of the wastewater ponds at the refinery also suffered damage due to shaking.

Refinery personnel indicated that soil improvement including micro piles, stone columns, and removal and replacement of soil soils were used at the site and that these areas of improved soil performed well at the refinery. In areas without improved soils, farther away from the main refinery, lateral spreading caused damage to petrochemical pipelines and large crude storage tanks (Fig. 2.39 and 2.40).

The Abastible natural gas plant and distribution center are adjacent to the ENAP refinery. The plant experienced minor ground cracking of fills, minor damage due to bearing soil strength loss, liquefaction, and lateral spread. The Abastible natural gas pier suffered minor damage during the earthquake; however, the uplift that occurred during the plate movement had decreased the draft such that ships can no longer reach the pier. No geotechnical damage to the major distribution facilities occurred.

Main transmission systems fared well during the shaking; however, localized areas within the natural gas distribution system suffered significant damage due to shaking and soft soils. Areas of the distribution system in Concepción will not be reconstructed due to the expense of repair. The tsunami also caused large damages to the distribution system. Further information about damage to natural gas and liquid fuel systems can be found in Chapter 5.

Figure 2.37. Damaged ENAP refinery intake lines at San Vincente

Figure 2.38. Stains on storage tank adjacent to ENAP refinery due to sloshing

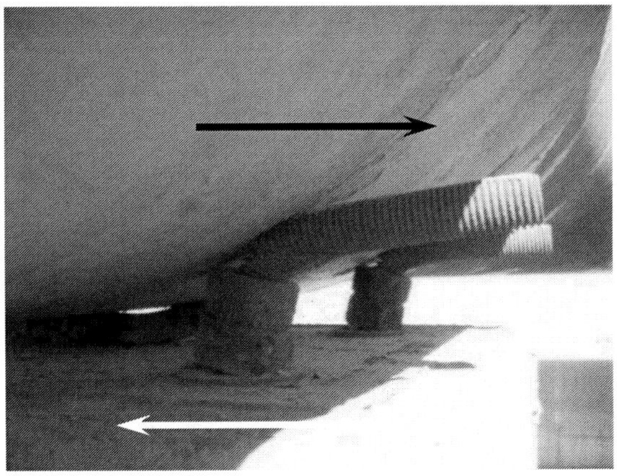

Figure 2.39a and 2.39b. Pipeline movement due to shaking and lateral spreading at the ENAP refinery

Figure 2.40. Lateral spread damage to storage tanks adjacent to the ENAP refinery. Separation can be seen between the concrete apron and the tank footing as well as in the piping in the background.

Colocation

It is common for several lifelines to be colocated especially at waterway crossings and along rivers. These waterway crossings are often located in areas of poor soils and areas vulnerable to lateral spread. When water crossings are damaged, several lifelines can be affected.

2.3 Discussion and Recommendations

Chile has a long history of earthquakes and is highly sophisticated in earthquake preparedness and earthquake engineering. Even still, each earthquake provides an opportunity to learn new lessons on how to better protect communities.

This earthquake is no exception. There is much that has been learned and much yet to be learned. Scientists and engineers are continuing research and improving their understanding of the seismo-tectonic setting, fault and earthquake parameters, fault slip and propagation, co-seismic land deformation including uplift, subsidence and lurching, ground motion attenuation, near field and site response, geotechnical ground effects, basin effects, topographic amplification, soil structure interaction, tsunami hazards, and more.

Much of the damage that the TCLEE team observed has also been observed in other earthquake settings. Three areas of concentrated damage are (1) areas of tsunami inundation, (2) areas of weak soils (i.e., slow shear wave velocity), and (3) weak infrastructure (i.e., structures that are not designed and constructed to resist lateral loads). In most instances including this earthquake, damage should have been expected in these areas.

1. **Tsunami inundation.** Many communities and lifeline infrastructure have been built in low-lying coastal areas that are vulnerable to tsunami inundation. Although it is possible in many places for people to evacuate safely to high ground above the tsunami flooding, the exposed infrastructure remains vulnerable to damage. Port facilities, unless protected by adequate wave energy dissipating structures, are inherently at risk.

2. **Weak soils.** Many communities and much lifeline infrastructure have been built on or near weak soils, such as dunes, flood plains, river mouths, and estuaries, and are especially vulnerable to seismically-induced ground effects. Co-seismic settlement, liquefaction, lateral spreading, landslides, basin effects, and soil amplification are common in weak soils. Infrastructure in areas of permanent ground deformation tend to incur extensive damage. Coastal highways on slopes, unconsolidated soils, and cut and fill slopes are often susceptible to ground failure. Embankments and bridge abutments commonly settle and move down slope. Wastewater outfalls and bridges over river crossings are common in areas with soil amplification, liquefaction, and lateral spreading potential and are inherently at risk.

3. **Weak infrastructure.** Many communities and lifeline infrastructure were built before the current day understanding of seismic hazards and are vulnerable during earthquake shaking. Older critical lifelines with important societal functions (such as

large power plants or refineries) with high consequences due to failures (such as dams or levees upstream of large populations) and with other sensitive parameters should be required to meet modern seismic levels and performance objectives. Additionally, lifelines that are colocated and/or interdependent with other lifelines should require special performance consideration.

The Maule earthquake was a significant earthquake not only for the affected public but also for professional earthquake community. In 1977 the U.S. Congress started a national earthquake program titled National Earthquake Hazards Reduction Program (NEHRP) to better protect society from damaging earthquakes (see www.nehrp.gov). Thirty some years since its inception, a successful NEHRP involves four government agencies with seismic experts, the science advisor to the U.S. president, and the federal budgeting office as well an advisory board from the professional earthquake community. Following the Maule earthquake, NEHRP has sponsored workshops to identified knowledge gaps and possible research activities (see www.eqclearinghouse.org/20100227-chile/research-needs-workshop).

Due to Chile's high seismic risk, TCLEE recommends that a Chilean national earthquake program similar to NEHRP with emphasis on research, preparedness, emergency response and recovery, and mitigation be established. Effective communication to the public, such as providing information on preparing emergency kits and public evacuation notices during tsunami emergencies, should also be included in the national program.

In addition to the development of a national earthquake program, the TCLEE team has the following recommendations:

- We understand that seismic considerations have been included in codes for buildings and certain other structures have existed for decades. We recommend that these codes be extended to lifeline engineering requirements and include soil design considerations. Seismic considerations should be included in design and construction of all lifeline systems.
- Location of lifelines as well as other critical structures like hospitals, police stations, and schools, should consider tsunami inundation areas, seismic hazards, landslides, and other geologic and geotechnical hazards. While some maps of geologic hazards, expected site amplification, and soft soils do exist, the mapping should be expanded and updated so that they can be consulted when new structures and lifelines are completed. These maps would be useful in areas like Iloca, where tsunami victims are being relocated to an area prone to landsliding. These maps should be included in building codes and regulations so that geologic and geotechnical considerations are included in planning and zoning efforts.
- Based on our observations of wide spread liquefaction and lateral spread damage, we recommend that soil improvement be used more widely to protect abutments, embankments, and other structures from seismically-induced ground failure. Replacement and recompaction is also a viable option; however, it is often difficult due to site proximity to waterways, depth of poor soils, and existing structures. It is also often very expensive. Many other soil improvement methods such as soil mixing,

grouting, stone column installation, and drainage installation can be more economical and easier to implement without large amounts of earthwork. New structures should consider using ground improvements; however, Chile should also consider improving poor soils in areas where critical structures and infrastructure are located. If ground improvements are not utilized, structures in areas of sensitive soils should be constructed on deep foundations that extend far enough into competent soils to resist seismic forces as well as any downdrag or lateral loads caused by liquefaction and soft soils. Improvement of bearing soils should also be included in the design and construction of retaining walls, especially at ports and other areas near waterways and on alluvial soils. Design of sheet pile and soldier piles walls in these areas should also consider liquefaction and lateral spreading loads.

2.4 Acknowledgements

We extend special thanks to the dedicated staff at SNGM and Ioan Nistor from the University of Ottawa, Canada. We also wish to thank Doug Bausch (FEMA), David Wald (USGS), and Kristin Marano (USGS) for PAGER information. Many people provided information and assistance with this chapter, including SNGM, numerous Chilean contacts, GHD, TCLEE team members, and more. The Ports, Water, and Bridges Departments of the Chilean Ministry of Public Works and the personnel at the Port of Lirquen (Pedro Gorizález G.) and Port of Coronel (Eduardo Faúndez R.) provided ASCE/TCLEE team with guided tours, photos, and relevant information and allowed us to use the photos and information in this monograph. We also would like to specifically thank Mauricio Villagran, Hugh Rudnik, the engineering department of the Universidad Catolica de la Santisima Concepción, Karla Gonzalez and Geri Pranzini and GHD Consulting, Christian Ledezma of the Pontifica Universidad Catolica de Chile, and Barbara Domont, and Nicolas Lira Turpaud.

Unless otherwise credited in the annotation of each figure and table in this chapter, all figures and tables belong to the ASCE/TCLEE team members.

2.5 References

Ben Brooks, James Foster, Mike Bevis, Bob Smalley, Hector Parra, Juan Carlos Baez Soto, Mauro Blanco, Eric Kendrick, Jeff Genrich, and Dana Caccamise, <http://supersites.earthobservations.org/chile.php>

Boroschek, R., P. Soto, and R. Leon, 2010. *Maule Region Earthquake, February 27, 2010, Mw = 8.8 – Renadic Report 10/08 REV. 1.* Civil Engineering Department, Faculty of Mathematics and Physical Sciences, University of Chile. October 2010.

CalTech Tectonics Observatory, 2010. California Institute of Technology <http://tectonics.caltech.edu/slip_history/2010_chile/index.html>

Earthquake Engineering Research Institute (EERI) Web site, <www.eqclearinghouse.org/20100227-chile>

Geo-Engineering Extreme Events Reconnaissance (GEER), <http://geerassociation.org/GEER_Post percent20EQ percent20Reports/Maule_Chile_2010 Ver2_Cover_Chile_2010.html>

Group on Earth Observations (GEO) Web site, <http://supersites.earthobservations.org/chile.php>

Hydrographic and Oceanographic Service of the Navy (SHOA) Web site, <www.shoa.cl/servicios/tsunami/tsunami_ing.php>

IRIS Web site, <www.iris.edu/hq>

Marcelo Farías, Gabriel Vargas, Andrés Tassara, Sébastien Carretier, Stéphane Baize, Daniel Melnick, Klaus Bataille' Land-Level Changes Produced by the M_w 8.8 2010 Chilean Earthquake *Science* 20 August 2010: Vol. 329. no. 5994, p. 916 DOI: 10.1126/science.1192094

MCEER Web site: earthquake engineer to extreme events, <http://mceer.buffalo.edu/research/Reconnaissance/Chile2-27-10/default.asp>

National Oceanic and Atmospheric Administration (NOAA) Web site, <http://nctr.pmel.noaa.gov/chile20100227>

Nistor, I., Palermo, D, Saatcioglu, M., Murty, T., (2010). Estimation of tsunami forces on Infrastructure, 9th US National and 10th Canadian Conf. on Earthquake Engineering, EERI, Toronto, Canada, 8 pp, CD-ROM.

Ruegg et al. (2009). Physics of Earth and Planetary Interiors. 175: 78-85. Vigny, C., A. Rudloff, J.C. Ruegg, R. Madariag, J. Campos, M. Alvarez.

Servicio Nacional de Geologica y Mineria Web site, www.sernageomin.cl/

UNAVCO Web site, <http://www.unavco.org/research_science/science_highlights/2010/M8.8-Chile.html>

U.S. Geological Survey, (poster), <ftp://hazards.cr.usgs.gov/maps/sigeqs/20070716/20070716.pdf>

U.S. Geological Survey Web site, <http://earthquake.usgs.gov/earthquakes/eqinthenews/2010/us2010tfan/finite_fault.php>

Wikipedia, <http://en.wikipedia.org/wiki/2010_Chile_earthquake>

3
TRANSPORTATION SYSTEMS

The transportation network includes highways, overpasses, bridges, and railways. The highways and bridges were the most seriously damaged in the magnitude 8.8 offshore Maule earthquake that occurred in Chile on February 27, 2010. There are approximately 12,000 bridges in Chile. Of the 100 bridges that were damaged, 20 collapsed (Fig. 3.1). The collapsed bridges include both pedestrian overcrossings and roadway bridges. The damage to the highway system was significant; many of the highway systems failed causing major disruptions in the movement of traffic in the metropolitan areas of Santiago, Constitution, Tubul, and Concepción. These systems should have been designed or retrofitted to a minimum performance level of no collapse (i.e., life safety).

Many of the bridges failed because of poor detailing of the connections and articulations. Proper attention was not given to following the load path from the superstructure through the bridge substructure down to the foundation with capacity protection provided through the load path. New bridges with skewed supports failed, and old bridges adjacent to the failed bridge with no skew remained standing illustrating the effect of providing restraint to resist the rotation of the superstructure in the horizontal plane of the bridge deck. Observations of failures illustrating the importance of following the load path to capacity protect the components using robust detailing so that the ductile components can absorb the energy.

Bridges also failed due to liquefaction and lateral ground-spreading effects, which occurred in Concepción with the Juan Pablo II Bridge crossing the Bio Bio River. Other bridges in Concepción connecting to the Juan Pablo Bridge also failed due to liquefaction.

Nearly 2000 of the 12,000 bridges in Chile are owned and operated by privatized concessions, contracted by the Chilean government using design, build, and operate (DBO) contracts. Bridges owned by these concessions were designed more recently and did not perform as well as those owned by the Chilean government. These concession-owned bridges are primarily located in the metropolitan areas and were the primary cause of the disruptions.

Figure 3.1. Collapsed bridges cause major disruption in metropolitan areas

3.1 Description of the System

There are approximately 12,000 bridges in Chile. Two thousand of these were designed by concession holders from other countries. The bridges designed by the Chilean Government and consultants under their control performed well compared to those designed by the Spanish concessioners. The Chilean government's bridge design practice, which began in 1960, was a combination of approaches used in Japan and the United States. Since 1971 its practice has been influenced by AASHTO seismic design criteria. In 1985 Chilean government practice was upgraded to more closely conform to the AASHTO 1983 *Guide Specifications for Seismic Design of Highway Bridges*, Division 1A, using an acceleration coefficient, $C_{S,}$ of 0.15 g, which was raised to 0.30 g in 2002. Beginning in 1996 the bridges designed by the Spanish concessioners were designed in accordance with Spanish bridge design standards. Bridges designed using these standards did very poorly in the earthquake.

3.2 Overview of System Performance

The damage to bridges designed and built after the mid-1990s suffered major damage compared to those built before the mid-1990s. The primary reason for the difference in performance is the change from the AASHTO criteria to the Spanish criteria and the corresponding change in construction details.

Most bridges in Chile are built with composite precast and prestressed girders with a reinforced concrete deck (Fig. 3.1). This typical bridge type is used for single or multiple span bridges. As illustrated in Figure 3.2, there are no intermediate diaphragms between the simply supported pre-stressed girders. Only vertical pipes are used to prevent uplift due to vertical accelerations. The girders are placed on concrete pads to provide the final grade elevations required for the concrete deck roadway surface. Transverse shear forces are resisted by metal brackets anchored in the concrete cap and fitted to the bottom flange of the precast girder as shown in the schematic diagrams in Figure 3.2. This poor connection detail and the absence of transverse shear keys and end diaphragms are the primary reasons for many of the bridge failures in this earthquake.

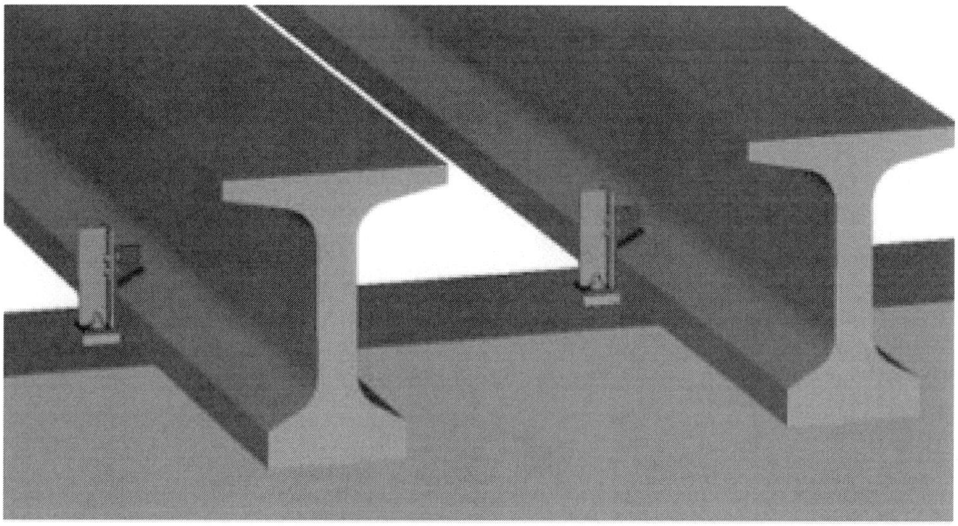

Figure 3.2a. Schematic showing connection detail as designed

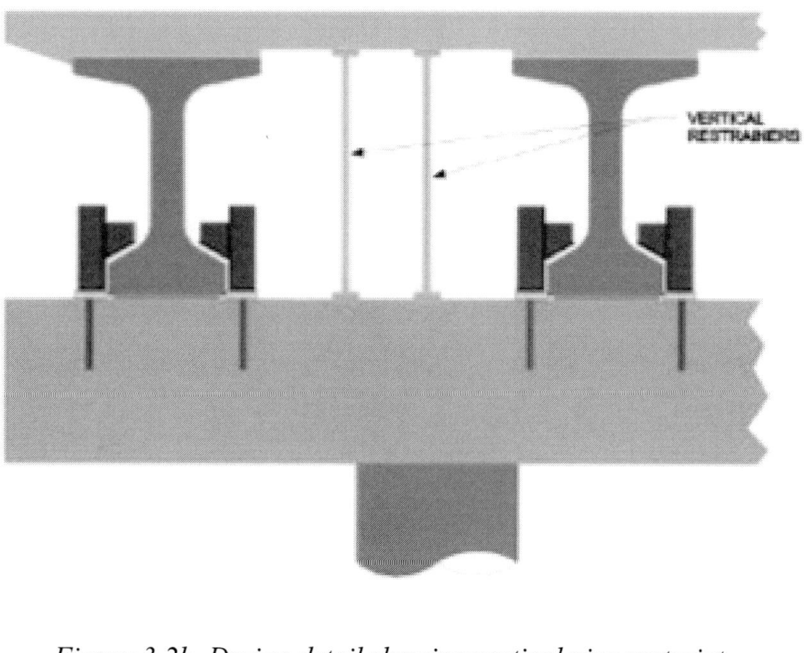

Figure 3.2b. Design detail showing vertical pipe restraint

The basic seismic design strategy used by AASHTO (2002, 2007, and 2009) transmits the seismic inertia forces generated at the superstructure level down to the foundation. This strategy was not used in Chile and is the primary reason for the poor bridge performance in the earthquake. The inertia forces generated in the superstructure are not transferred to the substructure through the weak connection. A simple analogy of a weak link in a chain is shown to demonstrate the connection's failure.

As shown in the force vs. displacement diagram in Figure 3.3, the basic seismic design strategy used in the most recent criteria AASHTO (2009) Guide specifications for load and resistance factor design (LRFD) of the bridge would have been to transmit the elastic superstructure inertia forces down to the ductile substructure following a selected load path. As shown in the figure, the weak link in the load path (i.e., the connection) did not have the capacity to transfer the load to the ductile column below.

The vertical restrainers (see Fig. 3.2b) are used to resist uplift. This practice began after the Santiago earthquake of March 3, 1985, which had high vertical accelerations. These vertical restrainers did not provide the transverse restraint that end diaphragms would providc. Elimination of the end diaphragms results in the loss of transverse and torsional restraint in the superstructure, which in turn contributes to the torsional strength of the overall system. The system as designed is torsionally weak, which contributes to the failure of the system as the torsional longitudinal modes of vibration cause an unequal distribution of the shear forces to the diaphragms.

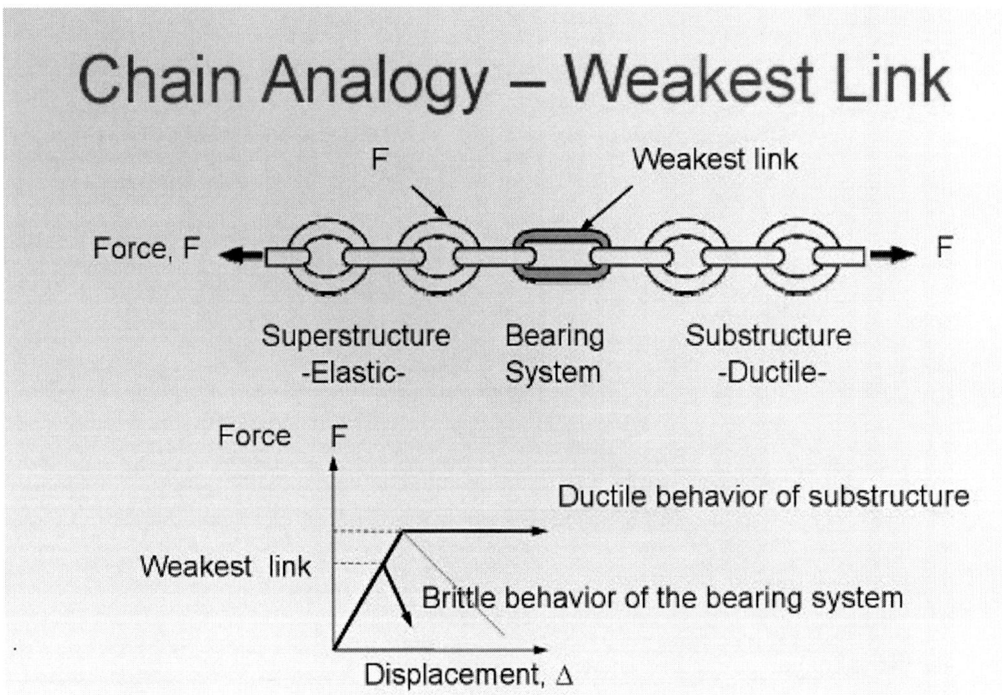

Figure 3.3. Schematic illustrating the weak connection between superstructure and substructure (Source: National Highway Institute 2011)

Before the mid-1990s, bridges were designed in accord with 1983 AASHTO specifications and would have had column shear capacities consistent with the plastic hinging capacity of the ductile columns. The connections should have been designed strong enough to transfer the superstructure forces to the ductile columns. Both the 1983 and the more recent 2009 AASHTO specifications require capacity design for all except the lowest of the design categories. Capacity design requires that all components have over-strength capacity to transmit forces to the selected ductile component or isolation system. Application of the new AASHTO specifications requires the designer to check the load path and design accordingly, which would have prevented all the failures that occurred in the Maule earthquake associated with deficient connection design for earthquake forces.

It is strongly recommended that Chile and its concessioners investigate the merits of the new AASHTO (2009) standards and require that all bridges for Chile be designed at a minimum for life safety (i.e., no collapse). Allowing seismic design procedures that are scaled down to minimum costs and rapid construction should not be permitted with the technology that has been available since 1990. The human suffering and economic loss witnessed in the disruption of the transportation system far exceed the minimal construction costs saved.

In addition, several bridges suffered significant damage or collapse during this earthquake due to liquefaction. This phenomenon has occurred in past earthquakes with follow-up research and development of methods to evaluate liquefaction potential. Design procedures have also been developed to reduce the risk of damage or collapse of a structure from liquefaction. Bridge damage due to liquefaction in this earthquake resulted from

1. lateral deformation of abutments and piers,
2. pile damage due to lateral spreading,
3. differential settlement at the abutments and piers, and
4. reduced foundation stiffness due to ground shaking and liquefaction.

A liquefaction hazard leading to damage or collapse can be reduced by considering ground improvement remediation of the soils at a bridge site and/or designing a bridge to accommodate the imposed displacements. Designing a new bridge or retrofitting an existing bridge may include either soil-site remediation and/or accommodating imposed structural displacements.

Figure 3.4 shows bridges designed before the mid-1990s that did not have a skew survived the earthquake standing adjacent to a post mid-1990 design that collapsed.

Figure 3.4. Skewed bridges—post-1990 collapsed bridge and pre-1990 undamaged bridge

3.3 Performance of Highways and Bridges

Bridges designed in accord with the Chilean practice did well compared to those designed with Spanish design practices by the concessioners.

The following list of observed factors contributed to the observed bridge damage and failures that occurred in the Maule earthquake:

- Bridges designed to be ductile with plastic hinging occurring at selected locations failed before the formation of plastic hinging, illustrating the presence of a weak link in the load path.
- Bridges collapsed because of inadequate support lengths and restraint.
- Bridges collapsed because of lateral spreading of the foundations and embankments at the piers and abutments.
- Shear connections failed at the intermediate piers and abutments.
- Skewed supports without proper shear connections and support widths either failed or moved to locations rendering the bridge unsafe to carry traffic loadings.
- Foundations failed due to liquefaction of foundation soils leading to support movements and rotations about the vertical axis of the bridge superstructure.

The following discussions were selected from the TCLEE team's reconnaissance investigations, information available through the media, and earlier reports from other reconnaissance teams. The locations selected include Santiago, selected portions of Route 5, Concepción, Tubul, and other locations where bridges or highways suffered notable damage.

Santiago

As shown in Figure 3.5, the roadway system in Santiago is laid out much the same as most North American cities. A network of roadways lead directly to downtown Santiago, which are connected by a beltway encircling the city. This system also contains several elevated roadway sections, which carry traffic in both directions.

Several roadway sections collapsed because of inadequate connections between the superstructure and substructure as shown in Figure 3.6.

Figures 3.6 and 3.7 show a failed connection, illustrating the inability of the designed shear restraint to resist the lateral inertia forces and transfer it from the superstructure down to the substructure, as shown in the weakest link analogy in Figure 3.3

Similar failures occurred at the abutments as shown in Figure 3.8. This abutment is on a skew causing an in plane rotation about the vertical axis of the bridge towards the acute corner of the bridge deck.

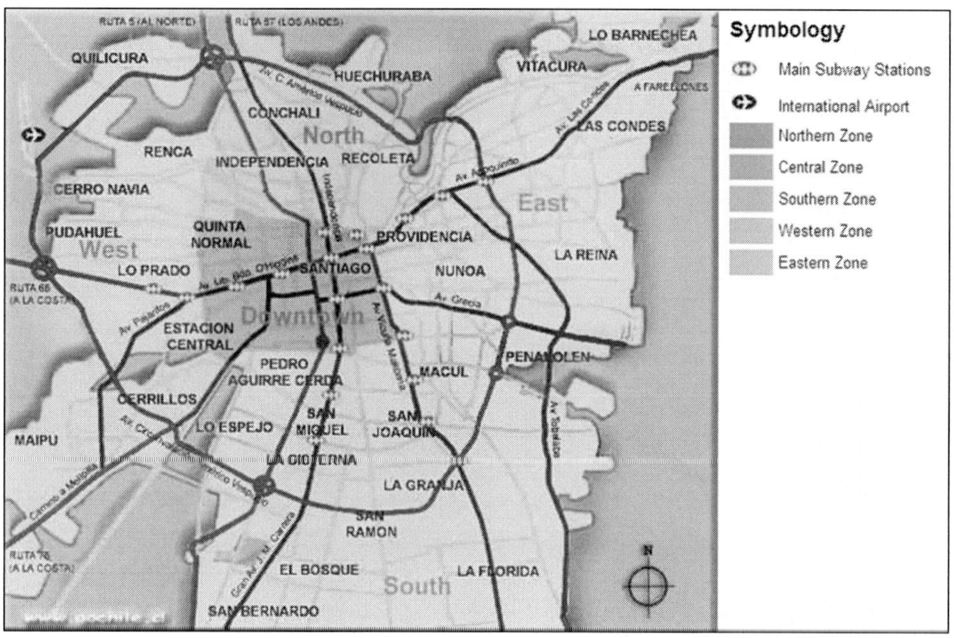

Figure 3.5. Santiago metropolitan area (Source: www.gochile.cl)

Figure 3.6. Collapsed span due to inadequate connection of superstructure to bent cap

Figure 3.7. Failed shear connection at the supporting bents

Figure 3.8. Failed connections at the abutments

Summary of Damage to Bridges and Roadways in the Santiago Metropolitan Area

The damage to bridges and roadways in the Santiago metropolitan area resulted from:

1. strong ground shaking and
2. design failures including the inadequate connections,
3. incorrect support lengths for the simple spans,
4. the omission of end diaphragms, and
5. longitudinal restraint between connecting spans.

Skewed supports without proper shear connections to prevent rotation of the superstructure about the vertical axis also contributed to the bridge failures. The current AASHTO specifications (2002 and 2007) and the most recent *AASHTO Guide Specifications for LRFD Seismic for Bridge Design* (2009) require that capacity protection be provided to ensure that there are no weak links in the load path from the superstructure down to the foundation. Current seismic design methodologies available for bridge designers should be incorporated into the bridge design practice in Chile.

Route 5

Route 5, the primary highway running north and south through Chile, suffered major damage due to bridge and roadway failures. A typical two-span overcrossing carries traffic across this main route (Fig. 3.9). These bridges are typically constructed of precast, prestressed concrete girders with a composite concrete deck. Each of the two spans provide for traffic movement in both directions spanning across Route 5. Several of the two-span bridges suffered damage, ranging from minor cracking to total collapse. The vulnerability of these simple two-span bridges is similar to the failures previously discussed.

Fig 3.9. Typical two-span overcrossing for Route 5.

The bridge shown in Figure 3.10 has no concrete end diaphragms between the girder and the girder ends supported on the bent cap. No transverse restraint is provided for the individual girders except for the restraint at the end of the bent cap for the exterior fascia girder. As shown in Figure 3.10, the transverse restraint has cracked. The shear restraint at the ends of the bent cap is inadequate to resist the transverse movement of the entire superstructure. Shear restraint is needed to transmit the superstructure inertia loads to the ductile columns to form plastic hinges in the column for a ductile design. The weak link in this case is the connection between the superstructure and substructure, as shown in Figure 3.3. The missing diaphragm is the most important thing. A bridge can survive if there is a diaphragm that can slide on the bent cap and avoid loss of support even if the shear key fails.

Without adequate transverse restraint, the superstructure will displace and eventually collapse as illustrated in Figure 3.11, which shows the transverse offset of the superstructure at the abutment. Typically, raised concrete shear keys cast between the girders are used to provide the restraint in the transverse direction at the interface between the superstructure and substructure to prevent this movement from occurring.

Summary of Damage on Route 5

The typical bridge over-crossings on Route 5 failed because of strong ground shaking and the weak connections between the superstructure and the substructure. The connections were inadequate to resist transverse movements and transmit the superstructure inertia force to the ductile columns of the substructure as described in Figure 3.3. This poor connection detail and the absence of transverse shear keys and end diaphragms (Fig. 3.2) are the primary reasons for many of the bridge failures on Route 5.

Figure 3.10. Inadequate transverse restraint at center support for two span overcrossing

Figure 3.11. Transverse offset of the superstructure at the abutment in Concepción

Concepcion

As shown in Figure 3.12, Route 150 crossing the Bio Bio River is a critical link connecting Concepcion and Talcahuano with San Pedro. As shown in the aerial view in Figure 3.13, there are three bridges connecting Concepcion and San Pedro. This critical link was seriously damaged with the collapse of a single approach span of the Juan Pablo Bridge II on the Concepción side of the Bio Bio River (Figure 3.14). A railroad bridge also crossing the Bio Bio was damaged, causing a disruption in the railway service between Concepcion and San Pedro.

Two of the bridges were being used before the earthquake; the third was out of service because it had been replaced by the Juan Pablo II Bridge.

The Juan Pablo II Bridge was completed and put into service in 1974. The 2,300 m long (7,580 ft.) bridge is composed of 70 simply supported 33.0 m (108.3 ft.) spans made of composite precast concrete girders with a concrete deck. This collapsed span was replaced with a Bailey bridge to temporarily reopen the bridge with limited traffic.

The bridge failed primarily due to liquefaction-induced lateral spreading of the embankment adjacent to the river (Figs. 3.15 and 3.16).

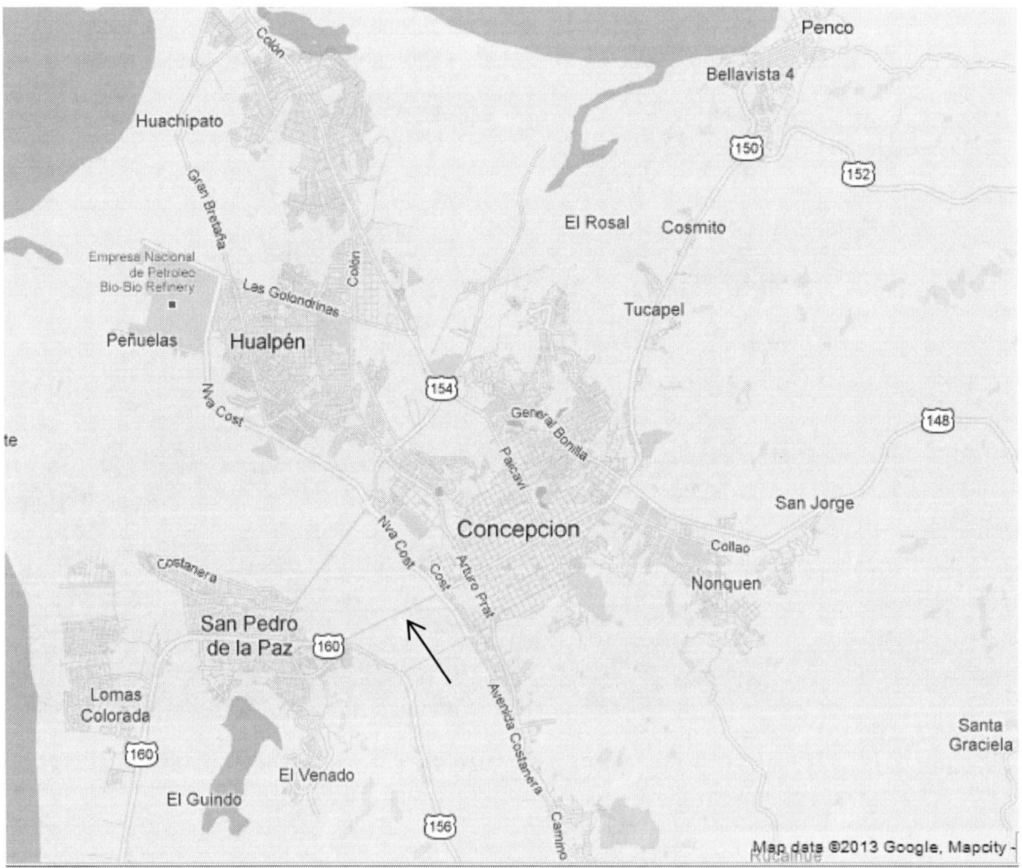

Figure 3.12. Route 150 crossing the Bio Bio River at Concepción (Source: Google Maps)

Figure 3.13. Areal View of the three bridges crossing the BioBio River at Concepción (Source: Image courtesy of the Image Science & Analysis Laboratory, NASA Johnson Space Center)

Figure 3.14. Collapsed span of the Juan Pablo II Bridge of the Concepción approach

Figure 3.15. Collapsed span of the Juan Pablo II induced by lateral spreading and liquefaction

The lateral movements of the supporting column bents in combination with the ground shaking resulted in the column failure and span dropping off the narrow hinge seat, similar to that shown in Figure 3.17. The closure for of the bridge for repair and the limitations on its use following the repair had a serious economic impact on bordering cities.

The railroad bridge crossing the Bio Bio River (Fig 3.18) was also damaged as a result of liquefaction.

Figure 19 shows the temporary shoring used to support a bridge damaged due to induced liquefaction.

Figure 3.20 shows the collapsed spans on the old Bio Bio River Bridge.

The two-span precast concrete girder bridge at Puente Juan Pablo II (Fig. 3.21) failed due to liquefaction resulting from settlement at the intermediate bent (Fig. 3.22). The bridge was located in a heavily traveled interchange, and its failure caused major disruptions with the closure of a route leading to the main crossing of the Bio Bio River. Liquefaction at this interchange was localized as evidenced by the failure of the roadways approaching this bridge (Fig. 3.2). Figure 3.24 shows sand boils adjacent to the bridge and its approaches.

Figure 3.16. Liquefaction-induced lateral spreading

Figure 3.17. Narrow hinges seat used on collapsed span

Figure 3.18. Failed Bio Bio River Railroad Bridge

Figure 3.19. Temporary support of the Bio Bio Railroad Bridge with timber shoring

Figure 3.20. Collapste spans on the Old Bio Bio River Bridge (Photo by Mark Yashinski)

Figure 3.21. Failed bridge on the north approach to the Juan Pablo II

Figure 3.22. Liquefaction failure resulting in pier settlement

Figure 3.23. Failed bridge approach roadway

Figure 3.24. Sand boils adjacent to the bridge provide evidence of liquefaction

Summary of Damage to Bridges and Roadways in the Concepción Area

The damage to bridges and roadways in the Concepción area was due primarily to strong ground shaking liquefaction-induced lateral spreading of adjacent embankments. A liquefaction hazard leading to damage or collapse can be reduced by considering ground improvement remediation of the soils at a site. Designing a new bridge or retrofitting an existing bridge may include either soil-site remediation and/or accommodating imposed structural displacements.

Also contributing to the bridge damage was the combination of the inadequate connections, the support lengths of the simple spans, and the omission of end diaphragms and longitudinal restraint between connecting spans. Current AASHTO specifications require that capacity protection be provided to ensure that there are no weak links in the load path from the superstructure down to the foundation. Current seismic design methodologies for bridge designers should be incorporated into the bridge design practice in Chile.

Damage to Bridges and Roadways on the Route to Tubul

Several bridges and roadways connecting to the coastal town of Tubul were seriously damaged. The Puente Tubul Bridge (Fig. 3.25) collapsed due to liquefaction. Support pier settlements and rotations led to the collapse. The narrow support widths with no longitudinal restraint between the spans contributed to the eventual collapse (Fig. 3.26). Roadways were also damaged due to liquefaction (Fig. 3.27).

Figure 3.25. Collapsed Puente Tubul Bridge (Photo by Mark Yashinsky)

Figure 3.26. Pier wall with narrow girder support seat (Photo by Mark Yashinsky)

Figure 3.27. Puente Tubul damaged bridge approach roadway (Photo by Mark Yashinsky)

Summary of Damage to Bridges and Roadways on the Route to Tubul

The damage to bridges and roadways en route to Tubul was due to strong ground shaking and liquefaction-induced lateral spreading of adjacent embankments. A liquefaction hazard leading to damage or collapse can be reduced by considering ground improvement remediation of surrounding soils. Designing a new bridge or retrofitting an existing bridge may include either soil-site remediation and/or accommodating imposed structural displacements

Also contributing to the bridge damage was the combination of the inadequate connections, the support lengths of the simple spans, and the omission of end diaphragms and longitudinal restraint between connecting spans. Current AASHTO specifications require that capacity protection be provided to ensure that there are no weak links in the load path from the superstructure down to the foundation. Current seismic design methodologies for bridge designers should be incorporated into the bridge design practice in Chile.

Figure 3.28. Collapsed unreinforced masonry arch bridge built in 1870 en route to Tubul Raqui

Other Locations Visited

The Silva Rodriquez Bridge crossing the Moule River (Fig. 3.29) suffered limited damage requiring that load limitations be placed on the bridge. The damage was located at the north abutment and included buckling of the end bearing stiffeners and fracture of the bottom flange and web. The composite steel girder concrete deck bridge is supported on steel columns with a chevron bracing system at the connections to the bent cap (Fig. 3.29). The sloped exterior combined with the center vertical column and the chevron bracing at the bent cap connection provides a very robust system to resist transverse earthquake loadings.

The more flexible superstructure moved in a transverse direction as indicated by the deformation of the brackets connecting them to the substructure (Fig. 3.31). This connection provided some limited restraint for the transverse seismic load, but in this case, it yielded to allow the superstructure to move on the bent cap to somewhat isolate the superstructure from the substructure. The design strategy of allowing the superstructure to move transversely on the bent cap is used both in retrofitting existing bridges and occasionally in new design.

Figure 3.29. Silva Rodriquez Moule River Bridge

Figure 3.30. Robust column to cap connection

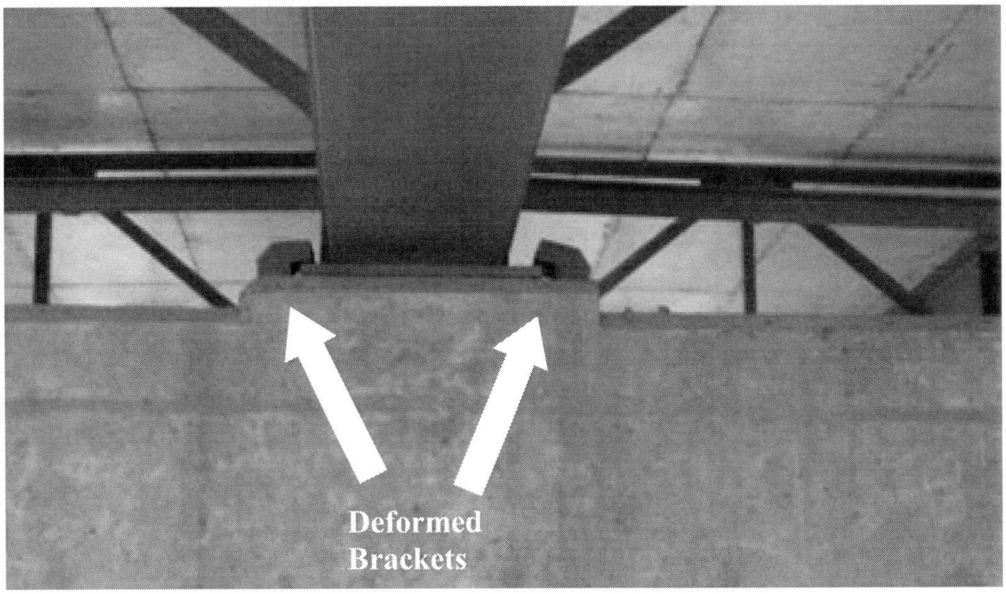

Figure 3.31. Yielded connection brakets due to transverse superstructure movement

A good strategy for seismic design is to allow movement of the superstructure at the connection to the substructure as shown in Figure 3.31. An isolation system would allow movement for the seismic event to decouple the superstructure from the substructure with an isolation bearing and return it to its original position following the earthquake. The isolation bearings provide a means of (1) decoupling the superstructure from the substructure, (2) taking energy out of the responding system, (3) providing damping thereby reducing the vibration effects, and (4) reducing the shear forces in the substructure.

A more recently recognized seismic design strategy is to absorb energy to allow deformations in the form of compression buckling and tension yielding to occur in the cross framing of the superstructure for this bridge (Fig. 3.32). This strategy is one of three now recognized in the recently developed AASHTO LRFD Seismic Design Guidelines for Bridges (2009).

After the earthquake the Mataguito River Pedestrian River Bridge between Licanten and Curepto (Fig. 3.33) was temporarily out of service. Unlike most suspension bridges, this bridge has a pinned connection in the tower to allow for rotation (Fig. 3.34).

As shown in Figure 3.34, the bridge was not damaged at the Licanten end. However on the Curepto side (Fig. 3.35), the bridge failed with the collapse of the approach spans. Note the suspension cables in the photo are undamaged and remain anchored in the original construction. The steel portion of the tower above the pinned connection remained plumb (Fig. 3.36). Note the temporary access to the bridge provided by the timber stairway. The concrete portions of the tower below the pinned connection moved off their foundation and rotated (Fig. 3.36 and 3.37). This failure is due to the lateral movement of the foundation materials due to liquefaction.

The pinned articulation used on this suspension bridge provided the redundancy needed to prevent the whole structure from collapsing.

Figure 3.32. Buckeled diagonals of end diaphragm cross bracing

Figure 3.33. Pedestrain suspension bridge in Curico

Figure 3.34. Pin connection of tower to pier providing for tower rotation

Figure 3.35. Collapsed approach spans

Figure 3.36. Plumb tower on rotated pier

Figure 3.37. Failed pier foundation

Figure 3.38. Bridge failure due the tsunami

Summary of Damage to Bridges and Roadways at Other Locations

The damage to bridges and roadways at other locations resulted from strong ground shaking, liquefaction-induced lateral spreading, and the tsunami. The strong ground shaking combined with inadequate connections, support lengths for the simple spans, and omission of end diaphragms and longitudinal restraint between connecting spans led to the bridge failures. Additionally, bridge damage was due to liquefaction-induced lateral spreading of the embankments at the abutments. Roadway damage was caused by liquefaction and the tsunami.

3.4 Railways

As shown in Figures 3.18 and 3.19 the railway bridge lies beneath one of the three bridges crossing the Bio Bio River at Concepción. The bridge's pier support on the Concepción approach was damaged due to liquefaction. As shown in Figure 3.19, the bridge is temporarily supported on timber shoring. The power for the railway system was damaged at several locations causing a major disruption in service. Figure 3.39 shows the railway bridge running parallel to Route 5.

Figure 3.39. Railway deck truss bridge parallel to Route 5

Summary of Damage to Railways

The observed damage to bridges on the railway system was induced by liquefaction and lateral spreading of embankments.

3.5 Observations and Recommendations

The damage to the transportation network had a severe impact on the public in the metropolitan areas. The damage was caused by bridge failures in the roadway and railway systems. The primary cause of these failures was inadequate design and structural detailing at the connections between the superstructure and the substructure, as demonstrated in Figure 3.3. These connections were detailed using vertical restrainer rods to prevent uplift caused by high vertical accelerations as experienced in the 1985 Chilean earthquake. This structural connection was not designed to resist horizontal earthquake forces, which was part of the weak link as shown in Figure 3.3. Failure to recognize that the 1985 earthquake was uncharacteristic of most earthquakes led to the design of the inadequate connections. Another factor contributing to the poor connection design is a violation of basic structural principles in providing a load path for the superstructure inertia forces to be transmitted down to the foundations. The failure of the foundation led to the collapse of the spans with narrow seats. In the absence of a properly designed connection, the ductile design approach of having plastic hinges form in the columns to dissipate energy was not achieved.

These connection failures can be avoided by following the design process described in the *AASHTO LRFD Guide Specifications for Seismic Design of Highways Bridges*. The new AASHTO guidelines place emphasis on designing and detailing connections to provide a robust connection, as needed, to transmit lateral earthquake forces from the superstructure through the connection to ductile substructure and foundation. The design procedure requirements for each of the four seismic design categories A through D (A being the lowest) are as shown in Table 3.1. A brief description of each category included in the design procedure is given below.

The design accelerations, SD1 for the design spectral value at 1.0 second periods for the four seismic design categories (SDC), are shown in Table 3.1. As shown, the design requirements vary depending on SDC. These requirements are described in detail in the new *AASHTO Guide Specifications for LRFO Seismic Bridge Design*. The first requirement to evaluate the liquefaction potential should be carried out in the early stages of the seismic design, so it can be considered throughout the seismic design process. The AASHTO guide specification is currently being presented to practicing engineers in the "ASCE Seminar on Seismic Design of Highway Bridges" by the 2007AASHTO guide specifications author.

As shown in the Table 3.1 the second requirement is the selection of a global strategy, which is recommended for SDC B and required for the two higher seismic design categories, SDC C and D. There are three global design strategies:

1. ductile substructure—elastic superstructure,
2. ductile superstructure—elastic substructure, and
3. elastic superstructure and substructure—with a fusing mechanism at the interface.

These three design strategies have been established from practice, observations from reconnaissance investigations, research and development, and the exchange of information between countries with earthquake hazards.

The requirement to identify an earthquake resisting system (ERS) involves identifying the load path to transfer the superstructure inertia forces through the bridge system to the foundation. Most of the failures in the Maule earthquake could have been avoided if the designers had complied with this requirement.

The support connection is required in all the SDCs. This requirement is critical in the transfer of the inertia forces and should not be a weak link in the load path as demonstrated in the sketch in Figure 3.3.

Support length is also required in all the SDCs. The design formulas included in the AASHTO guidelines are based on sound engineering principles and have been used for both new design and retrofitting of existing bridges. Inadequate support length has been a primary cause of failures to bridges in most of the earthquakes that have occurred to date, including the recent Maule earthquake in Chile.

Table 3.1. Design Requirements for the Various Seismic Design Categories

Requirements	Seismic Design Categories (SDC)			
	A	B	C	D
SD1	0.0 to 0.15	0.15 to 0.30	0.30 to 0.50	0.50 and above
Liquefaction Potential	—	Recommended	Required	Required
Global Strategy	—	Recommended	Required	Required
Identification of an earthquake resisting system (ERS)	—	Recommended	Required	Required
Support Connections	Required	Required	Required	Required
Support Length	Required	Required	Required	Required
Demand Analysis	—	Required	Required	Required
Implicit Capacity	—	Required	Required	—
Push Over Capacity	—	—	—	Required
Detailing - Ductility	—	SDC B	SDC C	SDC D
Capacity Protection	—	Recommended	Required	Required
P - Δ Effect	—	—	Required	Required
Minimum Lateral Strength	—	Required	Required	Required
Source: Data from AASHTO (2009).				

The demand analysis is a basic requirement in all aspects of bridge design for all types of loadings. There are special requirements for the seismic loadings to capture the effects of earthquake loadings, which are explained in the new AASHTO guidelines. Computer programs are readily available to bridge designers and, in many cases, have been tailored for use by bridge engineers.

The next requirement is the capacity determinations, which includes the use of implicit formulas that have been developed by conducting refined analysis on a large group of typically used columns. The analytical results are calibrated with the results of an extensive collection of experimental data on the behavior of reinforced concrete columns from the onset of cracking to failure. The developed formulas are included for SDC B and C. A push over is required for the highest seismic design category D. The capacity determination as described here replaces the previously used "R" factor used to estimate the force reduction for ductile behavior.

The ductility requirement is included to guide the designer in providing the amount of reinforcing steel and the location to achieve the ductility required to form the plastic hinge and provide the ductility needed to resist the earthquake forces.

The capacity protection requirement is the most critical task in ensuring that the transfer of forces to the ductile components is achieved in the design.

The P-Δ effect requirement is a design check on the displacements that occur as a result of the earthquake displacements to avoid instability due to excessive displacements.

The minimum lateral strength requirement is done at the final seismic design step to ensure that each vertical member has the ability to support its tributary load in an earthquake.

The AASHTO seismic design specifications described above are presented by the author as part of a seminar given by the ASCE. The seminar includes both bridge design and retrofitting.

3.6 Acknowledgments

This chapter was compiled with the assistance of practicing engineers from Chile, engineering staff from the Chilean Ministerio de Obras Publicas (MOP), and the American Society of Civil Engineers (ASCE) Technical Council on Lifeline Earthquake Engineering (TELEE) team participating in the investigation.

The professional engineers practicing in Chile that provided assistance in describing the damage and locating the damaged facilities included Professor Mouricia Villagran from the Catholica University in Concepción, Patricio Cordario, Geri Pranzini, Carlos Vega, Jorge Camnusano and Claudio Hernandez Toro from the GHD S.A., and Jaime Canales Memares representing the Departamento De Puentes on the repair of the Corico Pedestrian Suspension Bridge.

Bridge engineers from the MOP Bridge Design, Construction and Maintenance offices providing useful information, data and photographs on the damaged roadways and bridges include Andrew J. Tomlinason, Jose Antonio Noranjo, Jorge S.A. Vargas Bveni, Alejandro Molina Aguirre, Clauiso Catalso, and Carlos Sanhodeza Sanchez.

The ASCE TELEE team members all provided useful information in both group and individual discussion during the investigation. Both Allison M. Pyrch and Yumie Wang were very helpful in interpreting the Spanish, locating and driving to specific sites to investigate the roadway and bridge damage and to gather information from local residents. Special thanks to Alex Tang, out team leader, for the opportunity to participate in this TCLEE investigation and for his guidance and patience in preparing the report.

Financial support was provided by Victor Zayas, president of Earthquake Protection Systems, Inc. (EPS), located in Vallejo, California. Thanks also for editing, word processing, and graphic support to DeAnn Smeltzer, Carole Rouin and George C. Imbsen.

Unless otherwise stated in the caption of figures and tables, all figures and tables belong to the ASCE/TCLEE team members.

3.7 References

AASHTO, 1983. Guide Specifications for Seismic Design of Highway Bridges, Division 1A, American Association of State Highway and Transportation Officials, Washington, D.C.

AASHTO, 2002. Standard Specifications for Highway Bridges, 17th Edition, American Association of State Highway and Transportation Officials, Washington, D.C.

AASHTO, 2007. LRFD Bridge Design Specifications, American Association of State Highway and Transportation Officials, Washington, D.C.

AASHTO, 2009. Guide Specifications for LRFD Seismic Bridge Design, First Edition and interim revisions, American Association of State Highway and Transportation Officials, Washington, D.C.

National Highway Institute (NHI). 2011. *LRFD Seismic Analysis and Design of Bridges Reference Manual.* Reference manual for NHI Course 130093. Publication FHWA-NHI-11-030. Federal Highway Administration, Washington, DC.

For more information on the ASCE Seminar on Seismic Design of Bridges of Highway Bridges, visit *www.asce.org/conted/seminars*.

4
PORTS

Chile's economy is heavily dependent on ports, and the majority of the country's long coastline was affected by the earthquake and/or the tsunami. Some of Chile's largest industries were affected by the loss of port structures and stored cargo including agriculture, wine, fisheries, and wood products. Damage to inlet structures and off-loading stations impacted liquid fuel imports, fisheries, and other seaport-based industries. The ports within the influence zone of the earthquake were affected by strong shaking, weak and liquefiable soils prone to settlement and lateral spreading, uplift and subsidence, and the effects of the tsunami. The most devastating effects were caused by uplift in fishing ports, damage to structures due to liquefiable and weak soils, and tsunami wave action and flooding.

Significant damage was observed in port areas and to the industries dependent on the ports. Several large commercial ports were located within the earthquake zone including Quintero, Valpariso, Coronel, San Vincente, Talcahuano, and Lirquen. Several additional smaller fishing ports were also affected by the earthquake and tsunami. The Chilean Navy port at Talcahuano suffered significant damage. The ports observed by the TCLEE team included Talcahuano, San Vincente, Coronel, Lirquen, and the Abastible pier and other offshore structures in the Concepción area. These ports were located in areas of very gradually sloping shorelines and utilized offshore piers for the loading and unloading of cargoes. Many of the port storage areas and small vessel unloading areas were constructed with short (less than 3 m [10 ft.] exposed height) sheet piles and gravity block retaining walls along the waterline. Port infrastructure used by the fisheries industries was usually smaller, older, and often constructed with shallow timber piles. The larger, newer commercial ports were usually constructed with deeper steel piles and appeared to incorporate seismic design principles. The newly constructed pier at the Port of Coronel incorporated base isolation on the batter piles that supported the section of the pier under the gantry cranes. In general, the newer port structures performed well during the earthquake. Port damage is also discussed in Chapter 2 (Geology and Geotechnical). ASCE's Coasts, Oceans, Ports and Rivers Institute (COPRI) team also provided their findings at *www.asce.org/copri/COPRIHome.aspx*.

4.1 Overview of Ports Performance

The ports in Chile can be categorized into large commercial ports and small fishing ports, which are usually close to one another. The smaller and older ports and fisheries piers were hit hard by the earthquake and tsunami while the larger commercial ports experienced relatively moderate damage. A large portion of the damage in the waterfront areas occurred due to failures of walls that were founded on or in liquefiable and soft or weak soils that failed in either bearing or lateral movement during the earthquake. Container areas and storage areas as well as buildings within the lateral spread zone were affected by large-scale movements toward the water. These movements often extended 30 to 60 m (100 to 200 ft.) back from the water line, even in relatively shallow sloping areas. Lateral spreading and liquefaction were the main factors that damaged piers. The piers most heavily hit were the older piers where the foundations did not extend deep enough into competent material to resist lateral and down-drag forces caused by liquefaction, lateral spreading, and sediment accumulation. These piers were often used by the fisheries industry. The Port of Coronel had recently constructed a partially base isolated pier that performed very well during the quake.

Another effect of the earthquake that impacted ports performance were crustal rebound or lateral crustal movements, uplift, and subsidence along the coastline. According to the Ports Department of Chile Ministry of Public Works, the small fishing port of Lebu lost access to the majority of its port area due to upwards of 3 m (10 ft.) of uplift (Fig. 4.1). Figure 4.2 shows excavations to allow fishing vessel access were in action within days of the earthquake.

Figure 4.1. Lebu after uplift has lowered the sea level (Courtesy of the Ports Division of Chile Ministry of Public Works).

Figure 4.2. Excavations for a new access channel began in the Port of Lebu after uplift experienced during the earthquake (Courtesy of Ports Division of Chile Ministry of Public Works).

Commercial ports and manufacturing plants in port areas were hard hit by the tsunami. Loss or damage to goods in warehouses, ships, and near shore manufacturing and processing equipment due to flooding and wave action resulted in significant economic losses and loss of income to the nearby community. All industries that use the ports were affected by the earthquake, but based on information gathered during our reconnaissance, the fishing industry was likely the hardest hit due to the aging infrastructure and uplift in the coastal areas. We anticipate that it will take months to years for the fishery industry to recover the use of all of its ports and piers.

Commercial ports in general were working within a week of the earthquake to some capacity. Fishing ports were also working in limited capacities, even with severely damaged piers such as the one in Coronel (Fig. 4.3). However, the recovery of many of the ports operations visited by the TCLEE team was delayed by a lack of water and electricity, especially in the Concepción area where utilities suffered significant damage. Interdependence of lifelines definitely plays a critical role in the recovery process of any individual lifeline.

4.2 Coronel

The Coronel area we visited has two ports—a fishing boat unloading catch pier and a commercial port (Port of Coronel). The fishery pier, which is used for unloading catches, is about 500 m northwest of the Port of Coronel. Lots of fishing boats were moored in the bay adjacent to this pier.

Figure 4.3. Fishing boat pier used for unloading the daily catch was damaged by lateral spread at Coronel

Fishing Pier

The Coronel fishing port suffered significant damage due to lateral spreading to its main pier (Fig. 4.3). The crane at the end of the pier was not initially functional due to a severed power cable. Despite the extensive damage, with temporary power the pier and the crane were in use at the time of the TCLEE visit. Smaller boats were used to carry loads to the pier, and goods were offloaded by crane and hand carried to shore.

Figures 4.4 and 4.5 show details of two locations damaged by lateral spreading. The concrete block was rotated vertically close to 45° with the top pointing towards the shore. This created a gap of about 1.2 m (3.5 ft.) between the decks (Fig. 4.4). One of the piles next to this concrete block was torn from the pile cap due to differential movement of the pile and the deck (Fig. 4.5). The most probable reason for this type of failure is piles that were not driven deep enough into the more competent material. Additionally as this is not a commercial port, the pier may not have been designed and built to a higher seismic standard.

Power to the small crane was delivered by cables in a conduit mounted along the side of the pier, and cargo was carried to shore by hand. The power cable and conduit were severed when the decks moved apart (Fig. 4.6). Operations were extremely limited and the pier was expected to be replaced.

Figure 4.4. The deck support block rotated about 45□ creating a gap of 1.2 m

Figure 4.5. One of the piles was torn from the pile head by lateral spreading

Figure 4.6. Power cable attached to the side of the deck of this pier was severed

Port of Coronel

The commercial container port at Coronel had three piers. They are identified by numbers; the old pier is Pier #1, the new pier is Pier #2, and the conveyer pier that is under construction is Pier #3. Prior to the earthquake, Piers #1 and #2 were in use. The pier in construction suffered minor damage from earthquake shaking; however, the main structure was intact and construction had continued on the pier.

At the port entrance, the port's underground wastewater treatment plant had experienced upwards of 53 cm (21 in.) of uplift due to buoyancy forces in the liquefied soils (Fig. 4.7). Along the coast from the fishing pier to the Port of Coronel, extensive lateral spreading and resulting broken concrete pavement was prominent (Fig. 4.8). Further evidence of ground failure was noticed in the container area (Fig. 4.9). Strong shaking most likely caused a few top tier containers to fall. Additionally, underground services such as water pipes and power cables were affected by both lateral spreading and subsidence (Fig. 4.10). The main office building, a reinforced concrete structure had very minor non-structural damage; a small degree of subsidence was noted at the base of entry steps.

Figure 4.7. A concrete structure at the entrance of the Port of Coronel uplifted by liquefaction

Figure 4.8. Lateral spreading along the coast of Coronel

Figure 4.9. Extensive lateral spreading in the container storage area inside the port

Figure 4.10. Lifelines, such as this water pipe, were damaged by both subsidence and lateral spreading.

The first operational pier was built several decades ago and suffered severe damage due to lateral spreading. Figure 4.11 shows the damage to once vertical on shore piers. Figures 4.12 and 4.13 show a batter piles pair that was separated by up to 3 m and piles that have separated from the deck, respectively. Based on our observations, approximately one-quarter to one-third of the piers had been completely separated from the deck structure during strong shaking and ground displacement while the remaining piers were severely damaged. Although the pier deck did not show any significant damage (Fig. 4.14), the pier was no longer in use. A restoration plan had not yet been developed at the time of our discussion with the port engineers.

The second pier, which was recently constructed, was designed with seismic mitigation considerations. The piles were driven to depths of about 21 m (70 ft.) to reach the competent material and reduce damage due to lateral spreading according to design drawings shown to the team.

Figure 4.11. Lateral spread damage to piles at the Coronel older pier, Pier #1. The piles were originally vertical and were translated more than 10□ towards the water.

Figure 4.12. Batter pile pairs were separated up to 3 m by lateral spreading at Pier#1.

Figure 4.13. Pile separated from deck by lateral spreading at Pier #1. The weld between the pile and the plate broke apart.

Figure 4.14. Pier deck #1 showed no significant damage

In addition to the deep piles, the section under the deck of the massive loading cranes had platforms with base isolation bearings to reduce pile movement transferred to the deck and reduce probable damage to the cranes. These platforms are located along the centerline of the cranes (Fig. 4.15) and offset from the centerline of the pier by about 4 m (12 ft.). Figure 4.16 shows the platform with the isolation bearings. Figure 4.17 provides a close up view of the bearing, four bearings per platform. The top of the platform is about 8 by 8 m (24 by 24 ft.). The pier performed quite well. A few caps on the vertical piles along the pier perimeter showed small concrete cracks (Fig. 4.18).

Further, flexible utility connections, such as those shown in Figure 4.19 at the expansion joint of the main pier with the cranes, were part of the seismic design to reduce interruption of services to the pier operation. The bypass water pipe (circled in Figure 4.19) did not have a flexible connection. Fortunately, the differential motion did not severe the pipe, but it did pull the clamp from the concrete. This has to be corrected; however, the pier was back in operation as soon as electrical power was restored about a week after the earthquake.

Pier #3 of Port of Coronel was under construction during the earthquake and during our visit. This new pier is a conveyor pier intended for loading minerals and mining products. Figure 4.20 shows a partially completed pier.

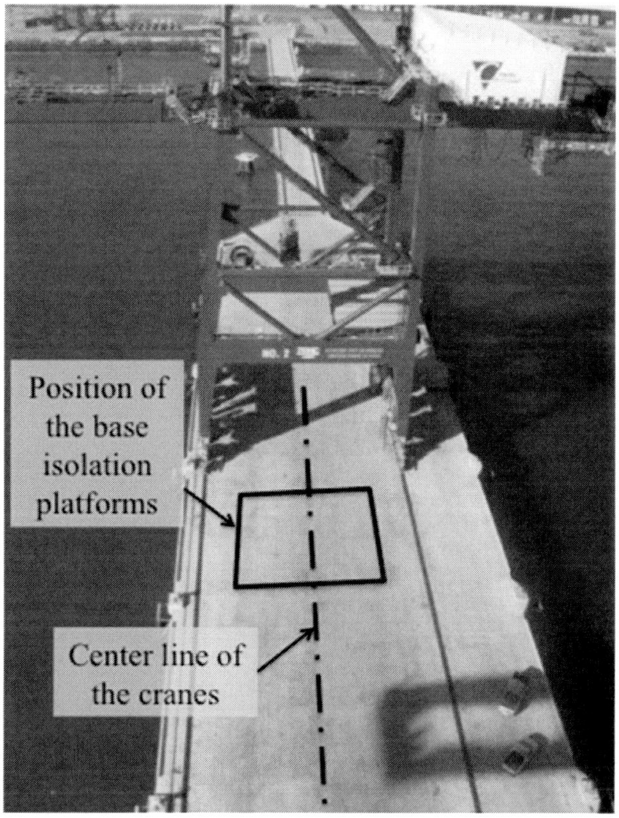

Figure 4.15. Location of the base isolation platform with respect the cranes

Figure 4.16. Rafts of base isolated batter piles in conjunction with non-isolated vertical piles of Pier #2 at Port of Coronel

Figure 4.17. Close up look at the base isolation bearing

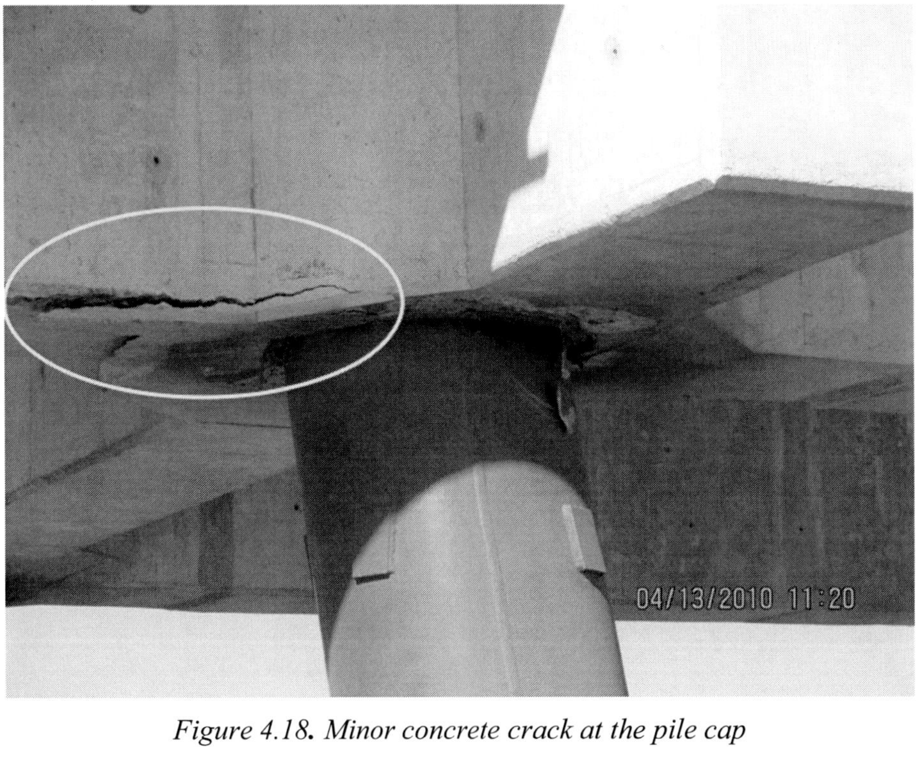

Figure 4.18. Minor concrete crack at the pile cap

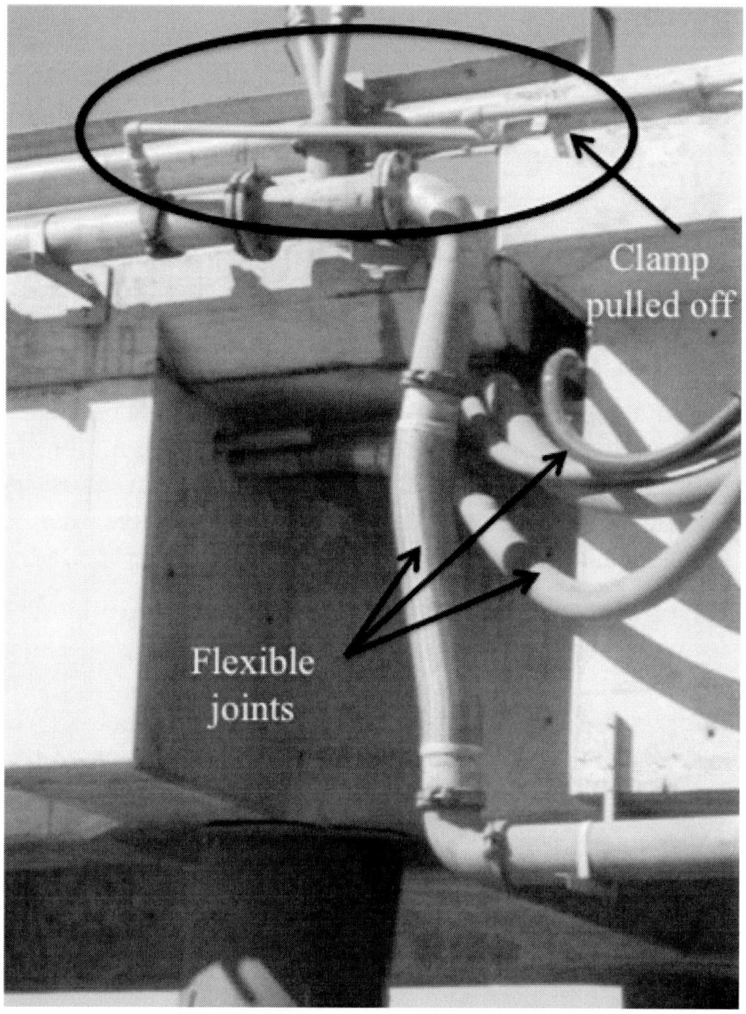

Figure 4.19. Flexible utility connections between the expansion joint on the newer commercial pier (Pier #2) at Port of Coronel are used to reduce differential movement damage.

Figure 4.20. Pier #3, the conveyor pier, under construction

4.3 Port of San Vincente

The TCLEE team observed several sections of the port area near San Vincente. This area includes fish process plants, oil intake landing site for the refinery, fish intake pipes and support structures, fuel tank farms, a fishery wharf, and a container port. Figure 4.21 shows the damage to the fishery wharf at San Vincente. The shallow retaining wall collapsed due to lateral spreading. Towards the end of the fishery wharf is the entrance to the container port of San Vincente. The TCLEE team visited this area on Sunday and was not able to tour inside the port. We observed that the containers were stacked not more than five high, which is the safe recommended stack height (Fig. 4.21). There was no evidence of any container having fallen from the stacks. As seen from the hilltop above the port, the wharf did not show any signs of damage (Fig. 4.22). According to the Ports Division of the Chile Ministry of Public Works, there was minor damage from the tsunami. It was reported that several containers were damaged and lost in the ocean. This area within greater San Vincente had more moderate damage than the beach areas opposite the tank farmers and the other areas where factories are located.

A few fish intake pipes connected to loading platforms a couple of hundred meters (about 600 ft.) in the bay were severely damaged due to both lateral spread and subsidence. Figure 4.23 shows one of the damaged fish intake structures; a large portion had sunken into the water. Temporary intake pipes were installed to unload the catch to the fish process plants across the street. The catch pumped through the pipes was discharged on trucks, which transported the fish to the nearby processing plants (Fig. 4.24).

Figure 4.21. Shallow retaining wall failure at the San Vincente fishing port. This fishing port is at the entrance of the Port of San Vincente.

Figure 4.22. Aerial view of the Port of San Vincente with significant damage observed

The two crude oil intake pipes to the ENAP refinery plant were broken at the landing site (Fig. 4.25 and 4.26). There was no evidence of oil spill along the beach; however, there was most likely still crude oil in the pipes. There was no report of damage to the off-shore loading platform. ENAP informed us that the landing structure would be completely rebuilt to meet current environmental standards.

Evidence showed that there was at least 2 to 3 m (6 to 9 ft.) of lateral spread and about 1 to 2 m (3 to 6 ft.) of subsidence at the beach area around the oil pipeline landing area.

There are two large fuel tank farmers located across the street from the beach area (Fig. 4.27). Some of the tanks have at least 4.5 million liters of capacity (Fig. 4.28). There were no signs of damage to any tanks within these two tank farms.

Additionally, a couple of new offshore off-loading facilities were being built in the area (Fig. 4.29). We were told that one of them was damaged by strong shaking.

One of the factories had a few buildings along the waterfront damaged due to lateral spreading (Fig. 4.30). There was no building collapse. Due to the electric power outage, most of the plants were not in full operation.

Figure 4.23. Damage to intake structures of the fish process plant at San Vincente

Figure 4.24. The loading area of the fish pumped through the pipes from the offshore platforms

Figure 4.25. The crude oil pipes landing site for ENAP refinery in Concepcion

Figure 4.26. Broken pipelines just behind the warning sign shown in Figure 4.25.

Figure 4.27. The tank farms in the San Vincente area

Figure 4.28. Fuel tanks contain 4.5 million LTS in this tank farm

Figure 4.29. Conveyor piers under construction, including one that was slightly damaged

Figure 4.30. A partially collapsed factory building towards to bay

4.4 Port of Lirquen

The Port of Lirquen has two operating piers. Both sustained minor damage but were back in operation within a couple of days. One is a conveyor pier for loading bulk products like fertilizer; the other is a container pier. Large warehouses are required because of the wood products handled by this port. Due to the increasing cargo volume, this port will be expanding its area by landfill towards the north part close to the entrance gate. We were informed that a new pier would be added. Figure 3.31 shows an aerial view of the port with key areas circled.

Strong shaking damaged two of the three main warehouse structures used for storage and staging. The two smaller warehouses close to the waterfront collapsed; the area had already been cleared when the TCLEE team arrived. The large warehouse opposite the collapsed warehouses had structural damage to the steel frame and trusses (Fig. 4.32). This warehouse stores wood products such as plywood.

At the back of the port, there are two more large warehouses with railway tracks through them for loading and unloading (Fig. 4.33). Most land shipments are delivered by train with tracks reaching to the back of the port (Fig. 4.34). Permanent ground deformation caused the concrete road surface to separate, which resulted in railway tracks embedded in the concrete snapping apart. The separation was about 50 to 120 mm (2 to 5 in.) (Fig. 4.35). Some of the separated tracks were repaired during the team's visit. A section of the track was cut to fit the gap; then the voids were filled with concrete to finish the repair (Fig. 4.36).

Figure 4.31. Aerial view of the Port of Lirquen

Figure 4.32. Damage to warehouse, which stores wood products, due to strong shaking within the port

Figure 4.33. Railway track in the warehouse at the back of the port for loading and unloading from the train

Figure 4.34. Train tracks at the back of the port for land shipments at Port of Lirquen

Figure 4.35. Snapped train tracks in Port of Lirquen

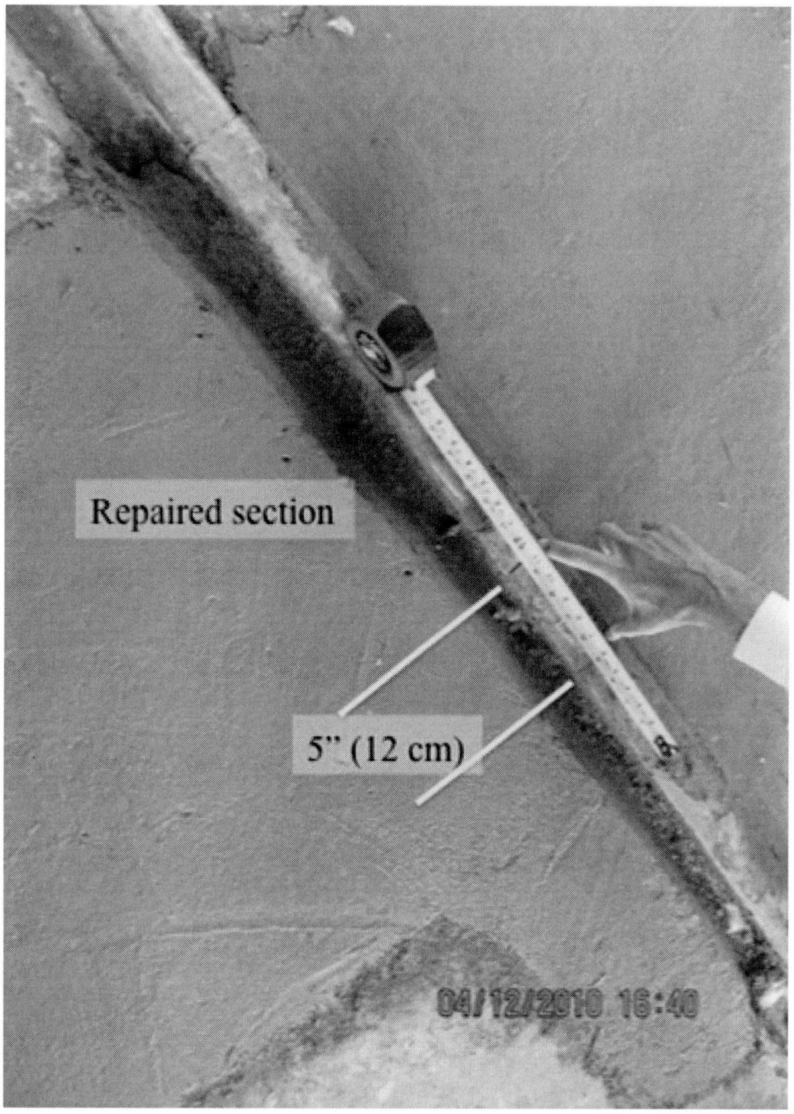

Figure 4.36. Repaired tracks at Port of Lirquen

4.5 Port of Talcahuano

The TCLEE team observed both the fisheries manufacturing and processing area (Isla Rocuant) and the container port at Talcahuano. Damage at the processing plants at Isla Rocuant due to strong shaking, liquefaction, lateral spread, or a combination of these earthquake-impact characteristics were not as evident due to the significant tsunami damage.

Talcahuano is essentially at ground zero of this earthquake. The community sustained extensive damage to buildings and lifelines. Many water leaks from broken pipes had not yet been fixed at the time of our visit.

Run-ups were up to 24 m (80 ft.) tall near the fisheries processing area. The cliff near this area as seen in Figure 4.37 reflects the water level reached by the tsunami. Flood marks on structures and buildings suggested that 2 to 3 m (6 to 9 ft.) of water were present for hours.

Fish Processing Facilities at Isla Rocuant

One of the fish processing plants closest to the water was destroyed, including all the equipment inside, and a few buildings collapsed. Figures 4.38 to 4.42 show the magnitude of destruction to the factories. Tsunami damage in this area had impaired several fish processing plants and materials storage areas.

All utility services in this area were interrupted. The distribution systems for electric power, gas, water, and telecommunication were destroyed (Figures 4.43 to 4.45). During the team's visit about 5 weeks after the earthquake, power was only partially restored. Factories that lost power used mobile power generators set to maintain operations (Fig. 4.46).

One of the fish storage tanks was moved by the tsunami about 200 m from its original position (Fig. 4.47) and was stopped by a small tree (Fig. 4.48).

Figure 4.37. Tsunami run-up level near Talcahuano reached about 24 m (80 ft.) above sea level

Figure 4.38. Damaged container, fish processing intake lines, and ships still present in the bay off fish processing area.

Figure 4.39. The tsunami destroyed one building and gutted another building at this fish processing plant at Isla Rocuant in Talcahuano.

Figure 4.40. Another tsunami damaged fish processing plant at Isla Rocuant in Talcahuano

Figure 4.41. Dislocated processing equipment inside the fish process plant. Note the supporting steel structure of the roof was knocked down by the heavy equipment.

Figure 4.42. Another equipment room destroyed by the tsunami

Figure 4.43. Electric power distribution poles were toppled by the tsunami, disrupting power supply to this area for a long time.

Figure 4.44. Telecommunication cable and distribution junction boxes were damaged when the utility pole was toppled.

Figure 4.45. A service man restoring the street lights to Isla Rocuant

Figure 4.46. Factories that were lightly damaged used portable power generators to maintain operation in Isla Rocuant.

Figure 4.47. The fish oil tank on the left had moved about 300 m from the concrete pad in the back (indicated by the arrow) to its present position.

Figure 4.48. The fish oil tank was stopped by the tree on the right, which prevented the tank from tipping over.

Container Port and Surrounding Areas

The commercial container port (Fig. 4.49) to the south of Isla Rocuant and on Talcahuano's downtown waterfront was also hit hard by the tsunami. News coverage and information from the Ports Division of the Chile Ministry of Public Works reported that containers from the port were thrown and piled five stories high against the adjacent apartment buildings by the tsunami. As shown in Figures 4.50 and 4.51, damage and residual debris were still present during the team investigation.

4.6 Navy Port at Molo

The Navy port at Molo near Talcahuano was also severely damaged by lateral spreading and strong shaking, according to the Ports Division of the Chile Ministry of Public Works. The soil loss due to lateral spreading, liquefaction, and likely helped by tsunami forces resulted in the retaining wall failure shown in Figure 4.52. Damage to buildings caused by strong shaking and tsunami is shown in Figure 4.53.

Figure 4.49. Container port on the waterfront of Talcahuano

Figure 4.50. Debris and tsunami damage to port structures at Talcahuano

Figure 4.51. Large ships were carried into town of Talcahuano by the tsunami

Figure 4.52. Damage to retaining walls due to soil loss at the Navy facility (Courtesy of the Port Section of Chile Ministry of Public Works).

Figure 4.53. Damage to structures due to strong shaking at Navy Port (Courtesy of the Port Section of Chile Ministry of Public Works).

Figure 4.54. The newer Abastible off-loading facility near Concepcion

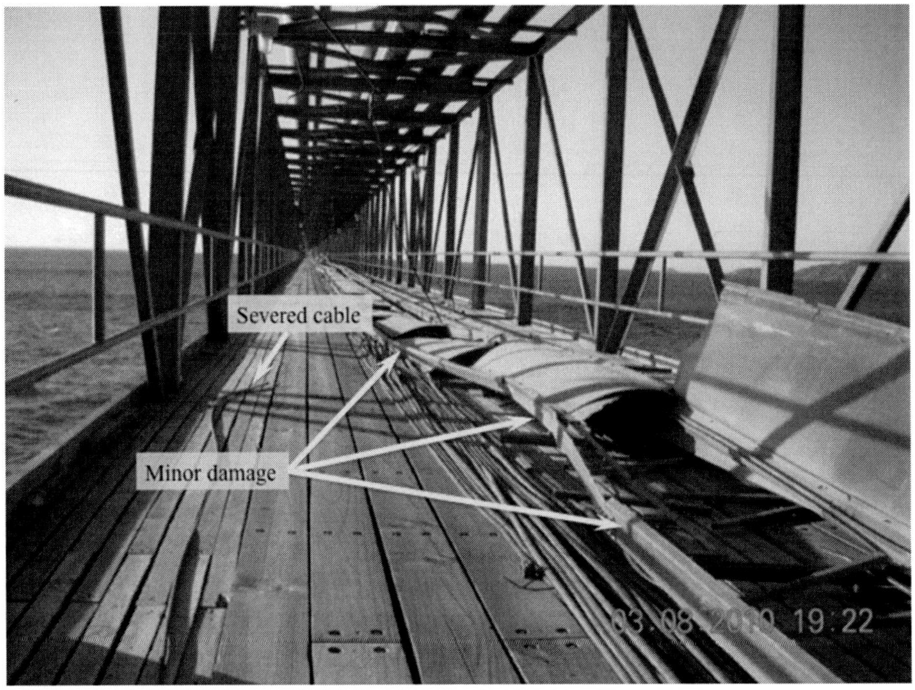

Figure 4.55. Minor damage due to strong shaking and uplift (Courtesy of Abastible)

4.7 Abastible Pier and Other Independent Offshore Piers in the Concepcion Area

The TCLEE team also observed a newer fuel off-loading facility near the ENAP refinery and Abastible facilities near Talcahuano. This pier performed reasonably well and experienced little damage with the exception of minor offsets and damage due to uplift, the extent of which was still to be determined by Abastible. The team was unable to secure design information on the pier, but we anticipate that the piles extend well beyond the liquefiable layers at the site and were designed to resist the large shaking loads. The pier is shown below in Figure 4.54. The damage to the pier is shown in Figure 4.55.

4.8 Major Observations and Recommendations

During our reconnaissance, we observed significant damage to the ports we visited and the areas around them. The following observations and recommendations are based on team observations and information collected during interviews port personnel.

- Significant damage was observed in areas of soft, weak, and liquefiable soils. Retaining walls and piers that were not designed to withstand lateral spread loads and/or were not founded on competent material suffered significant damage. We recommend that in port areas, resistance to liquefaction and lateral spread be considered in the design of these foundations. Ground improvement, deep foundations, or a combination of these methods should be used to support waterfront structures on competent materials.
- The partially base isolated pier at the Port of Coronel performed well during the quake. The elements of this design, such as flexible connections for utilities and

isolation of batter piles to reduce damage during differential shaking in conjunction with deep foundations driven through weak materials to competent soils, should be standard in all new construction.

- The major effects of the earthquake in port areas were the loss of infrastructure and cargoes due to tsunami damage. Goods stored at the ports were lost or damage due to the initial waves and subsequent flooding. Equipment and facilities were also damaged due to wave action and flooding. The fisheries industry was hit hard not only by the damage to piers and infrastructure but also by the loss or damage of many fishing vessels.
- Chilean ports experienced significant damage due to uplift during this earthquake. Several smaller ports such as Lebu lost freeboard that has severely limited or eliminated the ability to bring in large vessels for off-loading. Uplift in the Concepción and Talcahuano areas effected several piers including the newer Abastible LPG off-loading pier.
- Based on the damage observed in ports areas, and the significant economic impact of the damage, we recommend that seismic vulnerability assessments shall be an urgent project of all major and economically significant ports in order to develop a recovery plan after a damaging seismic events. Further, we recommend that procedures such as those required by California's Marine Oil Terminal Engineering and Maintenance Standards (MOTEMS) be adopted for all major ports and in water facilities. This document is available at www.slc.ca.gov/division_pages/mfd/motems/motems_home_page.html

4.9 Acknowledgements

The authors are extremely grateful to the following individuals and organization that provided valuable information and allowed us to use photographs of damage in this chapter.

Chile Ministry of Public Works – Ports Section

ENAP (Refinery in Concepcion)

Abastible (Producer of LPG)

Ms. Karla Gonzalez and Mr. Geri Pranzini of GHD, S. A.

Mr. Pedro Gorizález G. of Port of Lirquen

Mr. Eduardo Faúndez R. of Port of Coronel, and

Professor Mauricio Villagran of Catholic University of Concepcion, Engineering Department.

Unless otherwise specified, all figures and tables belong to the team members of ASCE/TCLEE investigation team.

5
GAS AND LIQUID FUEL

The overall performance of the gas and liquid fuel systems in the earthquake-impacted areas was quite good. BioBio refinery had experienced minor setbacks due to a collapsed heat reflector in the heater room. The refinery was shut down for about 4 months because the main crude oil feeder pipelines were damaged. Instead of repairing the pipelines, the whole landing area was redesigned and rebuilt to satisfy environmental measures.

There was no damage reported to the Aconcagua refinery, located in Region V (Region de Valparaiso) on the coast northwest of Santiago. It was closed for about 1 week to allow a complete safety inspection before restarting operation.

5.1 Description of System

There are three petroleum refineries in Chile: Aconcagua in the north and generally west of Santiago, Bio-Bio near Concepcion, and Magallanes, a much smaller refinery well to the south in Chile. All refineries belong to ENAP (Empresa Nacional de Petrolio).

The Aconcagua refinery has a capacity of approximately 100,000 barrels per day (bpd); the Bio-Bio refinery has a capacity of 130,000 bpd. The distance from the epicenter to the Aconcagua refinery is approximately 330 km and about 130 km to the Bio-Bio refinery.

Crude oil is imported to the refineries. Gas is imported from Argentina via pipelines through the Andes.

5.2 Refinery Performance

Both of these two larger refineries were shut down by the earthquake. Electric power supply failures and damage appraisal were the principal causes of the initial curtailment of refinery operations. Damage at the Aconcagua refinery was minor, and the refinery was up and operating in about a week. The shaking level at Vina del Mar (closer to the epicenter) was about 0.34 g. Damage at the Bio-Bio refinery was not considered extensive, but what damage did occur was to critical elements. Specifically, there was damage to the heaters. Refractory (brickwork) is an essential component of heaters and is not rigidly connected to the structural frame of the heater; this refractory worked loose and fell to the heater floor during the long and intense ground shaking. Repair of the heaters entailed the removal and replacement of all of the refractory and insulation where utilized. Also on vertical heaters, which generally have a height to diameter ratio of 4 or 8 to 1, with a cone shape section at the base and a smaller diameter stack to the top of the heater, the base of the smaller diameter stack failed due to buckling.

Figure 5.1. Oil sloshed and spilled from the air bleed holes along the top of the tank.

There was spalling of concrete on some equipment foundations and some stretching of anchor rods; neither condition appeared extensive. There was also some minor local buckling of structural columns, but again not extensive.

Also at the Bio-Bio refinery, there was extensive sloshing of liquid in the tanks. Where liquid level was high, there was sloshing over and outside the tanks. There were three 275,000 barrel (bbl) floating roof tanks 67 m by 14.6 m (220 ft. by 48 ft.), two of which were partially full. Sloshing outside the tanks occurred at these two tanks (Fig. 5.1). In the tank farm area of the refinery, some permanent movement of larger pipelines (12 in. and above) on their concrete supports-movements up to a foot were also observable.

In the wastewater treatment area, there was soil spreading under the lined sloping sides of the retention basins. In the American Petroleum Institute (API) oily water separators, there were some scrapers, which jumped their guide rails. There was also damage to a cooling tower. At the coker unit, which receives the heavy residual liquid oil from other units in the refinery process train and turns this residual into solid petroleum coke, there is a 25 m (80+ ft.) drill tower, which sets atop a 25-m code drum structure. Some of the bolts between the drill tower and drum structure sheared, with a slight rotation of the drill tower structure. The base plates of the drill tower were welded to the top of the drum structure to ensure stability and continuity between the two structures (Fig. 5.2).

Figure 5.2. Bolts sheared between the first and second stages of the coker and drill tower, with slight rotation of the tower.

The Bio-Bio refinery was constructed in the 1990s and benefitted from the current state of earthquake engineering. Considering the local shaking of 0.65 g H and 0.58 g V, coupled with the long period of strong shaking (more than 2 minutes), the refinery performed quite well.

Products from both major refineries are distributed by trucks directly from the refineries and by pipelines. Pipelines emanate from both refineries and proceed toward central Chile but do not meet. Inquiries produced no indication of damage to either pipelines or terminals.

5.3 Crude Oil Pipeline System Performance

Except for the damage of the pipeline at the landing site by San Vincente Port (Fig. 5.3), the land portion of the feeder pipeline had minor damage of a few pipe supports due to ground deformation. During the earthquake there was no crude oil pumping through the pipelines; therefore, the pipe break at the landing site did not cause any oil spills in the bay.

Figure 5.4 shows the pipelines that feed into the refinery. Lateral spreading was observed along the transmission pipeline system. That caused some minor damage along the pipeline, but the steel pipe was not broken.

Figure 5.3. Landing site of the crude oil pipes

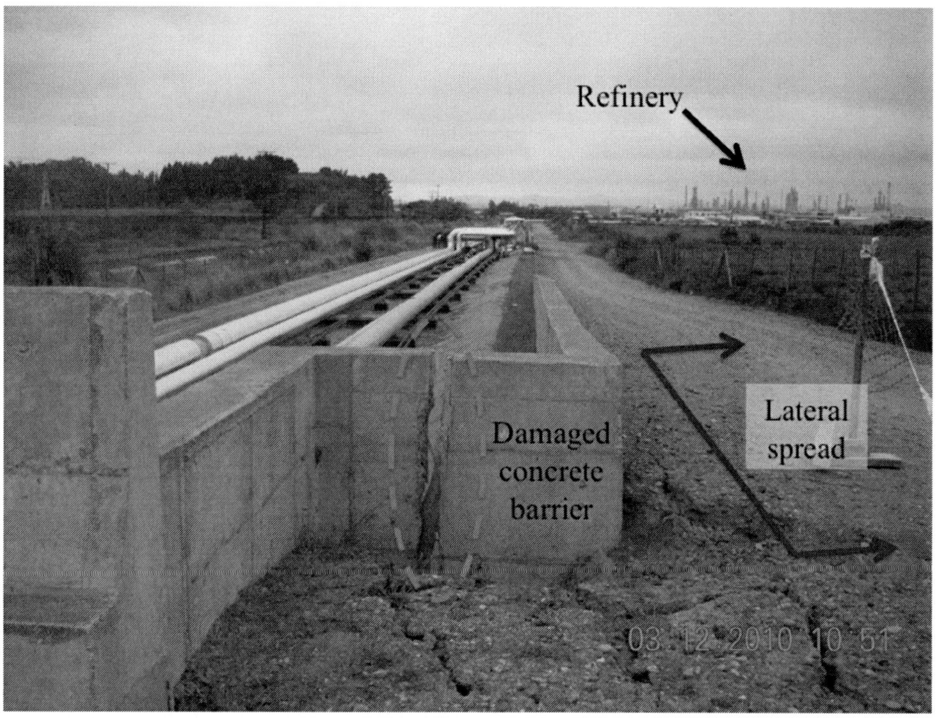

Figure 5.4. Oil pipelines that transport oil to the refinery

Figures 5.5 to 5.8 show the damage of the support saddles along the pipeline. All the observed damage mainly was due to subsidence of the concrete support due to ground deformation and liquefaction.

Figure 5.5. Subsidence of the support block

Figure 5.6. Pipe support was lifted off the block

Figure 5.7. Support block subsidence and movement along the pipe

Figure 5.8. Lateral spreading around the walk bridge over the pipes

5.4 Liquid Petroleum Gas (LPG) Plant Performance

Adjacent and to the west of the Bio-Bio refinery is the Abastible LPG plant; both sites are constructed on similar soil. The Abastible site was cleared of vegetation, and 3 m of select compacted fill was imported and placed on it. The plant, including two 15-m diameter LPG storage spheres, was constructed on top of this. This facility incurred no earthquake damage.

The three large tanks (275,000 bbl) noted above are located between the Abastible facility and the Bio-Bio refinery. These tanks are API 650 tanks constructed in the early 1990s, basically on the same soil conditions noted above. Lateral spreading occurred adjacent and to the west of one of these tanks, which was about two-thirds full. The spreading encompassed a length of about 30 m (Fig. 5.9). On this west side of the tank, the bottom plate pulled away from the bottom annular ring and allowed oil leakage in a small but clearly visible area.

5.5 Gas Fuel System Performance

Santiago is supplied with fuel gas from an LNG terminal on the coast. From inquiries, there was apparently no significant damage to the system or interruption of service. Santiago's outlying areas are supplied with LPG by local distribution.

Figure 5.9. Lateral spreading (or soil settlement) is shown at the center of the picture.

Gas fuel supply to Concepción and the remaining countryside is principally LPG. Abastible is a producer and supplier of LPG, but not a distributor. Central Concepción used to receive and utilize natural gas from Argentina; however, this is no longer a significant source of supply. Overall, the gas fuel system worked quite well in this earthquake. The extensive damage at the Port of Talcahuano was largely caused by sunami; however, cast iron pipe systems collapsed due to extensive ground settlement in the area.

The Bio-Bio refinery has extensive on-site storage of butane and propane in long horizontal storage vessels. There was apparently some minor concrete spalling at the anchor bolts with some anchor bolt displacement. Local distribution is via smaller pressure containers. Again, inquiries suggest that service may have been delayed but not interrupted by the earthquake due to both telecommunication interruption and road damage.

5.6 Major Observations and Recommendations

The Bio-Bio refinery was constructed in the late 1990s and benefitted from reasonably good seismic engineering design. Ground motions at the specific site are not known, but the long period (close to 60 seconds) of strong shaking certainly tested the facilities. The overall seismic performance of the refinery is considered good. The long-period sloshing, even at a great distance from the epicenter, must be considered in tank design to reduce spillage and tank damage.

5.7 Acknowledgments

The authors appreciate the assistance of Mr. Jaime Inzunza Vera, Jefe Ingenieria y Servicios of Abastible, and Mr. Fernando Raiqueo Vaccaro, Division Ingenieria of ENAP Refinerias (BioBio). Their valuable information and guided tour of their facilities during our visit provided us with in-depth knowledge of what happened during and immediately after the earthquake.

Both continued communicating with us via e-mail, allowing us to obtain the most current recovery and upgrades of their facilities. We are extremely grateful for their extensive time and effort to help us to clarify the contents of this chapter.

6
ELECTRIC POWER SYSTEMS

The transmission network performed well. There was limited equipment damage, and the networks' transmission substations were energized within 24 hours of the main shock. The distribution system busses were energized within 1 to 3 days, and most customers' power was restored within 2 weeks. There was a relatively high failure rate for transformer bushings, which is significant because of their potential impact on network performance. The long, narrow configuration of the system dictated by the shape of the country and the topography of the land limits transmission line route dispersion and system redundancy. While Chile uses much of the same equipment as the western United States, it makes extensive use of pantograph and semi-pantograph disconnect switches and candlestick live-tank circuit breakers at higher voltages. For lower voltages, it uses older air-blast circuit breakers, which are used sparingly in the western United States. Eight semi-pantograph disconnect switches,14 candlestick live-tank circuit breakers, and four air-blast circuit breakers failed, but these failures represented a small percentage of inventory in contrast to the failure of 29 transformer bushings. This count primarily reflects the sites visited.

The backbone 220 kV and 500 kV systems had been designed and installed during the past 25 years with earthquake provisions, and with a few exceptions, seismic practices, such as equipment anchorage and slack in conductor connections to equipment, were done very well. Some high-voltage substations make extensive use of damping devices on equipment, exclusive of power transformers. Damping devices are rarely used in the United States. An important thermoelectric generating facility near Concepción was damaged and out of service for several months forcing changes in the dispatch of power. Damage to equipment operating at 115 kV and below is less common in the United States. The low-voltage (34.5 KV and below) distribution system sustained minor damage due to collapsed buildings and damaged poles. The damage to concrete distribution poles was not significant, although poles are more heavily loaded in Chile than in the United States. Nonetheless, it was much more common than observed with the wooden poles used in the United States. Damage to wooden distribution poles in the United States is quite rare, except as secondary damage or due to soil failure. In the areas struck by the tsunami, the power infrastructure was destroyed, although damage to distribution poles was probably from debris carried by the tsunami rather than water loads.

Some power companies use cell phones to dispatch repair crews rather than the radio systems used in the western United States. Unfortunately, the loss of power disrupted cell phone usage and adversely affected the initial recovery of power systems. The duration of strong shaking in this subduction earthquake was much longer than has been experienced to date in the contiguous United States. The damage produced does not suggest that the long duration of shaking adversely affected most substation equipment, although it might have affected transformer bushings. Peak ground acceleration is usually used to characterize the seismic exposure of the equipment at the substation. No substations had strong motion instruments to record ground motions. The response spectra of the strong motion measurements had very high spectral accelerations characterizing the seismic exposure at a site by its peak acceleration, which may underestimate the motions that affect substation equipment. Although this chapter does not provide ground motion estimates at sites, it does describe the response of office equipment often used to estimate ground motions. The investigation team had unprecedented access to facilities, and there was open disclosure of damage data by utility personnel.

6.1 System Description

The power system in Chile is the product of the shape of the country, its topography, the population distribution, the evolution of its ownership, and system control. Chile is a very long and narrow country, and much of the population and ground infrastructure is located in the center of the county. This is true for the power system infrastructure as well. Chile has two principal grids—one in the north, which was not affected by the earthquake, and the main grid in the center and south. The power system is generally a linear system running along the length of the county. Hence, the diversity of transmission line routes is limited as well as the redundancy inherent in a power grid that covers a land expanse this length and width. The 500 kV backbone system on the West Coast of the United States has a similar form, although its substantial 230 kV system provides route diversity and redundancy. Most references to the United States refer to the western part of the country where seismic awareness and seismic hazard is relatively high.

The current configuration of the power system in Chile is also a product of the corporate evolution of the power industry in the country. Starting in the late 1980s, the system has been evolving from essentially a government monopoly to a segmented system that is divided into separate generating, transmission, and distribution entities. Transelec and TRANSNET (formerly CGE Transmission) play major roles in the high-voltage transmission function, subsidiaries of CGE (Conafe, Chilectra and CGE Distribution) play major roles in the distribution function, and ENDESA, Colbún, and AES-Gener are major players in generation in the main grid. In the western United States, there is a diversity of power systems with some large vertically integrated systems, some large transmission systems, an array of independent generating sources, and large and small municipal distribution companies, some of which are vertically integrated. In Chile, regulatory laws define transmission and distribution systems. Assets outside of substations operating at 23 kV and below are distribution systems, and those above 23 kV are transmission systems.

Power dispatch is the control of the output of power sources (thermoelectric and hydroelectric plants) so that an independent system operator (ISO) balances power supply to meet demand. Typically this is done by economic dispatch, which selects the most cost-efficient source to meet the demand, consistent with the capacities of the transmission network. This is easier said than done because generation includes low-cost hydroelectric power derived from water from the Andes, which can be seasonal and significantly influenced by years of drought, and more costly thermoelectric power derived from coal, gas, and oil derived largely from foreign fuel sources. In recent years the selection of the lowest cost source of power (from the short-term perspective) can and has resulted in severe seasonal power shortages because thermoelectric sources cannot meet needs without some hydroelectric power. Complicating the issues is that the natural gas used by many thermoelectric sources was cut off from outside of the country. This history has resulted in the common use of emergency power in many facilities, a situation not observed in the United States. For example, tall buildings with elevators, even residential buildings, often have emergency generators to operate elevators and provide other critical power needs, although this is not mandated by law.

A major source of hydroelectric power is from the Laja Lake, located east of Concepción, although much of the population is centered in Santiago. Mining and other industrial facilities are also major power consumers. The major non-hydro source of power (120 MW) in the Concepción area dedicated to providing power to the grid is the Bocamina power plant near Coronel. Additionally, Petropower's 60 MW cogenerating plant sells surplus power to the grid. There is also a large expansion of thermoelectric generating capacity in the region. Several plants in Chile were powered by gas from Argentina, but unilateral cuts of gas transfers have motivated diversity to other sources such as liquefied natural gas (LNG).

6.2 System Characteristics and Response

The following sections will contrast the power system configuration in Chile to that in the western United States. General comparisons will be made about the types of equipment, installation practices, and earthquake performance of equipment.

Independent System Operator (CDEC-SIC) and Power System Communications

The independent system operator (ISO) monitors the system and controls power source output to balance demand. Monitoring and dispatch is done with a supervisory control and data acquisition (SCADA) system. Typically, economic dispatch is used. That is, power demands are met by increasing generation of the most efficient generators. In Chile a key criteria is to conserve water so that the country is not totally dependent on thermoelectric sources in drought years because it cannot currently meet demand. It should be noted that several large thermoelectric plants are being constructed in the south so that power shortages triggered by severe drought can be avoided in the future.

The SCADA system in Chile is physically much different from that used in the North America. In Chile, SCADA communications use a carrier system, that is, high frequency signals are superimposed on power transmission circuits. This is used for voice communications, monitoring system status, protective relaying, and controlling generator output, although the latter may also use landline telephone circuits. In most parts of the transmission grid, microwave operates in parallel with the carrier system. The circuits with carrier signals have wave traps on the lines where they enter the substation to remove the signal before it enters the switchyard. A capacitive coupled voltage transformer (CVT) is connected to the line on the line side of the wave trap to bring the data into the control house for interpretation and action. While standards have been developed for the SCADA system, a combination of SCADA systems are used by each of the independent companies that have been integrated under deregulation. In addition, there has been a move to eliminate personnel from substations, which has placed an added burden on the SCADA system for substation alarms and control. The SCADA system used by ISO is almost 10 years old, and its capacity has been overloaded providing information that is not always correct. For example, the SCADA system may indicate the state of a circuit breaker different from its physical state. This makes it difficult for the ISO to properly control the system. This deficiency was recognized, and the ISO was in the early stage of planning a new SCADA system at the time of the earthquake. The installation of the new SCADA system will now be expedited. Some of the dispatchers also used landlines; however, after the earthquake these communication channels were also disrupted, which complicated dispatch. As power system elements came back online after the earthquake, some parts of the system came up independent of other parts. Isolated parts are called islands.

The ISO does not have its own alternate operations facility for dispatch. Should its facility be rendered inoperable, arrangements have been made for the ISO to use the major transmission company's operations center.

Individual utilities use the SCADA system for communication, which may be supplemented with landlines. Utilities generally use wireless systems (cell phones) to dispatch service crews to repair the network for both transmission or distribution systems, but transmission companies have an extended VHF radio-based system. As noted in the sections on telecommunication systems and interdependencies, cell phone service was generally disrupted after the earthquake, and this made it difficult to assess system damage and to dispatch crews to make repairs. In addition, as microwave repeaters were connected to distribution lines for power after the earthquake, some repeaters went off line once the battery back-up power system ran out of charge.

In the United States, there are many differences in utility communication systems. Carrier systems are being phased out and have not recently been used for voice communication, although they could be. Most utilities own a microwave system, which is connected to most transmission substations. These communication assets are usually used for voice communications and SCADA operations. The carrier system is primarily used for protective relaying, that is, the control of circuit breakers for system protection. Because most utilities like redundant communication channels, optical fiber is now widely used, thus eliminating the need for the carrier system. Many of the utilities in active seismic zones on the U.S. West Coast also have large internal communication systems. Each utility typically has a single private branch exchange (PBX) that acts like a small central office of a telephone company. In some cases they may not be that small; PBXs can accommodate several thousand lines. The PBX also serves as the main hub for most communications, including the utilities' microwave and fiber systems as well as the connection to the public switch network (landline system). In this sense the system vulnerability has been increased because there is no redundant PBX and a common failure mode can impact most communication assets. Although the PBX will be provided with emergency backup power, the seismic installation of some systems is wanting. In Chile the main transmission company has designed its phone network based on several PBXs taking in account the possibility of losing at least one.

Most U.S. utilities have their own radio systems that are used to dispatch service crews. Radios are also used for this purpose in power plants. Usually a centralized call center receives reports from customers about problems, and the call center sends a message to the appropriate maintenance center. The maintenance center uses a base radio station to dispatch crews, which are usually located at the center along with a store of spare parts. Because of the size and terrain of the service area of many utilities, repeater stations are needed. This adds to the cost and vulnerability of the system because, in an earthquake, power is often disrupted to the repeater stations, which are often located on remote hilltops, and access to provide fuel to backup power can be difficult. While wireless phones may be used, most use a radio system.

The California Independent System Operator (CISO) serves the same function as the ISO in Chile. The primary dispatch center is located in Northern California, but there is an alternate backup site in Southern California. CISO has several communication systems for contacting generating facilities. CISO has a contract with a major telecommunication company to provide communications, and contact is maintained to generating sites. In addition, power company communication assets, including microwave and optical fiber systems, are used. Dispatch is usually checked every 5 minutes, and updates are sent to generating units every 15 minutes. As in Chile, economic dispatch is usually used, but under certain atmospheric conditions, a pollution dispatch is used.

There are some differences in network reliability methods between Chile and the United States. The so-called "N-1" system requires that if any single facility, such as a generating unit at a power station or the transmission lines on a single right-of-way, is disrupted, there should be no disruption of service. In Chile this system generally applies to transmission lines, but not to all components. For example, on March 14 one of two transformers at the Charrua substation tripped off line. It was reported that a control circuit was damaged by the earthquake. This transferred the entire load to the remaining transformer, overloading it and causing it to trip off line. This resulted in a cascade of disruptions that affected 94 percent of the population of the entire country. Service was restored within a few hours, however. After the earthquake, five sensitive areas were identified, and at the time of our visit, this number had been reduced to three. For the Charrua substation, this required less than optimal economic dispatch to limit power flow through the substation to the capacity of just one of the transformers.

In the United States, a private, voluntary reliability system, called the North American Electric Reliability Corporation (NERC) was instituted after the massive 1965 Northeast blackout that affected several states and parts of Canada. As a result, reliability councils were established for the United States and parts of Canada and Mexico, and the N-1 system and other procedures were instituted. Notwithstanding these procedures, massive blackouts affecting several states in the Midwest and western United States have been triggered by minor disruptions, often because reliability guidelines were not followed. The Western Electric Coordinating Council (WECC), which covers the area from Denver, Colorado, to the West Coast and includes parts of Mexico, and Canada, is one of eight regional reliability entities.

After a system blackout, such as the one triggered immediately after the earthquake, there can be difficulties in bringing the network back up as an integrated system, particularly in a system as large as Chile's main interconnected system, which extends some 2254 km (1400 miles) from north to south. Isolated sections, called islands, can be energized. The difficulty is in connecting the islands. For two islands to be connected, they must be operating at the same frequency, the power in each island must be in phase, and the capacity of the link or links that connect the islands must have sufficient capacity to carry the necessary synchronizing power. An integrated network is always a balancing act to keep supply equal to demand. If demand is slightly higher than the supply, generators slow down, thus the system frequency is continually fluctuating slightly from its target value. If demand in one exceeds capacity when two islands are connected , the power to meet the demand will be drawn over the connection. On February 27 there were four islands; on February 27-28, three islands; and on February 28, two islands. The complete system was integrated on March 1 and 2.

For example in the United States, WECC, which has more than 1600 utilities, is not synchronized with the rest of the country because the connections between it and outside systems cannot carry the synchronizing power. DC links, which do not need to be synchronized, are used to connect WECC to other systems. For a "black start,"[1] a source of power is needed to start large thermoelectric plants; this is usually done with a hydroelectric source or a large (25 MW or more) peaking unit. In the United States, hydroelectric sources are often not available, and for various reasons the peaking units have not worked as desired for black starts.

Differences in Equipment

This monograph uses some terminology differently that normally used by the power industry. For some equipment, such as circuit breakers and disconnect switches, all three phases are considered a unit by the industry. At higher voltages these items are actually composed of three physically separated pieces of equipment, one for each phase. In this monograph when citing damage, individual phases are referred to if they are stand-alone equipment items. Thus, the failure of two circuit breakers refers to two phases; they could be part of the same circuit breaker group or different circuit breaker groups if each is an isolated item of equipment. Some items are constructed with all three phases in a single unit. In this case the item will be referred to as a three-phase circuit breaker. Many items within a switchyard, such as instrument transformers, wave traps and surge arresters, have always been referred to as isolated items because that is how they are constructed and used in the system.

Seven types of equipment or installation methods found in Chile are different from those typically found in the western United States. These include pantograph disconnect switches, horizontal pantograph disconnect switches, candlestick circuit breakers, air-blast circuit breakers, bulk-oil dead-tank circuit breakers, grounding switches, and damping devices to enhance the seismic performance of equipment. In some cases the difference is a matter of degree. Differences will generally be noted in the text.

Figure 6.1 shows a 220 kV pantograph disconnect switch (DS), none of which was damaged. Figures 6.2 and 6.3 show 500 kV semi-pantograph disconnect switches in the closed and open positions, respectively. One of the concerns about pantograph switch performance during earthquakes is that they connect two different structures. The switch is attached to the ground and the connection point is an overhead cable (Fig. 6.1 and 6.3) or a dead end structure (Fig. 6.2). Thus, the switch must accommodate the relative

---

[1] the process of restoring a power station to operation without relying on the external electric power transmission network

deflections introduced by the earthquake response of the different structures. While the live part of the switch is not massive, it has a high center of gravity so that it can place large moments on the supporting post insulators. These switches are occasionally found in the United States, but one substation in Chile had more than 60, 500 kV pantograph switches. Figure 6.2 shows a semi-pantograph switch with the upper contact (referred to as the trapeze in this report) supported from a post insulator that is suspended from a dead-end structure. All of the damaged pantograph switches we observed had this configuration, and the failures occurred in the connections at the bottom of the arms. In Figure 6.3, this occurred in the lower arrow points to the retracted arm and the upper arrow points to the trapeze, which in this case is connected to an overhead conductor.

Figure 6.4 shows a 500 kV disconnect not typically used in the western United States, which is referred to in this report as a horizontal semi-pantograph disconnect switch. It has the appearance of a center-break disconnect switch when closed, except that there are three support columns rather than two.

Figure 6.1. 220 kV pantograph disconnect switch

Figure 6.2. Closed semi-pantograph DS

Figure 6.3. Open 500 kV semi-pantograph DS

Utilities in Chile make extensive use of live-tank circuit breakers, which currently are rarely used in the western United States, where dead-tank circuit breakers are typically used. A live-tank circuit breaker gets its name because the unit contains elements that break the circuit at line voltage. With this design the bushings connected to the assembly can be much smaller, but the assembly must he isolated from the ground. This requires that one or more insulators support the live parts. Because of the high center of gravity, the insulators must resist large seismic loads, so these circuit breakers tend to be more seismically vulnerable. Early high-voltage versions of these devices introduced in the United States in the 1970s or before proved to be seismically vulnerable. Figure 6.5 shows a 500 kV candlestick circuit breaker in which three interrupter heads are used. Note the generous slack in conductor between the heads. Dampers have been installed at the base of each column (Fig. 6.6), which allows the column to rock, so generous slack is required. While less generous slack was provided between the circuit breaker and the pantograph disconnect switch, there was no damage. This unit was installed at the Charrua substation, located southeast of Concepción. Several candlestick circuit breakers failed, but none failed at 500 kV. Of the 13 sites visited or seen through gates, only one bulk-oil dead-tank circuit breaker was observed. While they are being phased out in the United States, there still are many in service.

Figure 6.4. 500 kV horizontal semi-pantograph disconnect switch

Figure 6.7 shows a 220 kV air-blast circuit breaker. The seismic performance of this unit has been enhanced as shown in Figure 6.14. Air-blast circuit breakers in the western United States, which did not have seismic enhancement, have largely been replaced due to poor seismic performance.

Figure 6.8 shows a grounding switch used to ground a bus within the substation when personnel perform maintenance. In the United States disconnect switches often have a grounding blade that is used to separate a transmission line from a substation. The grounding blade is used to ground the line rather than leave it floating to pick up stray charges. In the United States grounding switches are typically not found in the switchyard for maintenance reasons.

Chile uses damping devices in substation equipment more extensively than the United States to improve earthquake performance. Many of these devices also modify the equipment's' natural frequency, thus providing a type of base isolation. These devices were primarily used on 500 kV equipment, and their use varied depending on the substation. In the Polpaico Substation, north and west of Santiago, only two types of

Figure 6.5. 500 kV candlestick live-tank circuit breaker

equipment used supplemental damping. Figure 6.9 shows dampers on an instrument transformer, and Figure 6.10 shows base isolated live-tank circuit breakers. This figure shows a pendulum supported, live-tank circuit breaker with friction damping. Note that this 500 kV circuit breaker has a single head. Other circuit breakers in this substation had multiple heads with the same damping configuration. This support system has been used on more massive circuit breakers and was also observed in the switchyard at the Manzanillo generating station in 1995 earthquake in Mexico, but has not been seen in the United States. Figure 6.11 shows the friction damper used on the circuit breaker, and the dust marks indicate the extent of motion during the earthquake. These marks could be used to estimate the magnitude of the ground motions if system parameters were known. Figures 6.12 and 6.13 show the dampers on a current transformer (CT) and potential transformer (PT), respectively. Clearly a variety of damping devices have been used. It is reported that almost all types of 500 kV and 220 kV equipment in Transelec substations have been shake table tested.

Figure 6.6. Dampers at the base of each circuit break column

Figure 6.7. 220 kV air-blast circuit breaker

Figure 6.8. Bus grounding switch in substation

Figure 6.9. Arrow at right points to damper on base of equipment

Figure 6.10. Suspended live-tank circuit breaker with supplemental damping

Figure 6.11. Friction damper on circuit breaker with indication of motion

Figure 6.12. Dampers on CT

Figure 6.13. Dampers on PT

Figure 6.14. Belleville washers add damping to CB column

Figure 6.14 shows Belleville washers that have been affixed to the base of the head support column on an air-blast circuit breaker to enhance its seismic performance by adding damping and changing the as-installed frequency of the equipment.

6.3 Generating Systems

Table 6.1 summarizes the damage observed or described to the team during a visit to the Bocamina generating stations and to the Petropower cogenerating station, which was not visited. The Bocamina substation adjacent to the power plant is also discussed.

A visit to the Bocamina coal-fired generating station (170 MW) near Coronel south of Concepción, which was constructed in 1970, had similar problems to those experienced by the Moss Landing Power Plant in the 1989 Loma Prieta earthquake. Figure 6.15 shows an overall view of the boiler support structure and stack. This boiler, like most,

Table 6.1. Summary of Damage to Generating Facilities

Bocamina
Concrete stack damaged at construction joint
Steel stack liner damaged
Onc 2" boiler tube broken
2 out of 20 rods suspending the boiler damaged
Seismic stops or dampers damaged
Six pipe hangers damaged
Generator foundation settled about ¼ inch (6 mm)
Crane in powerhouse had to be realigned
One control room infill wall failed
Some control room suspended ceiling tiles fell
Construction joint between the boiler support structure and powerhouse moved and was slightly offset
Tubes surrounding generator low voltage conductor opened
Water supply was disrupted
Coal conveyor system in coal yard damaged
Unit to be out of service for several months
Petro-power
Stack damaged and unit out of service

is hung from the top of the boiler support structure by spring-supported rods. Two of the 20 rods failed. There is no cladding surrounding the support structure to enclose the boiler combustion chamber, so there must be lateral restraints to limit lateral deflections from wind and earthquake loads. In California these are usually called seismic stops and they are typically sacrificial elements. There may have also been dampers to limit lateral motion. Some of the dampers and/or seismic stops were damaged. The seismic stops limit lateral motion but allow vertical movement, which is needed due to thermal expansion when the boiler is heated and cooled. The most serious damage was to the stack. This repair will take several months. Figure 6.16 shows some spalled concrete at a construction joint. The steel liner was also damaged, and the character or severity of this damage is unknown. Typically, the connections between the liner and the outer stack are damaged, and this is a simple repair.

Figure 6.15. Overview of stack, boiler, and boiler support structure

A 2-in. boiler tube in the boiler broke. This is a simple repair, but the boiler must be allowed to cool, so the repair crew can enter it. Figure 6.17 shows an enlarged view. The lower arrow points to a pipe hanger, six of which were damaged. The upper arrow on the right points to what some boiler manufacturers call a buck stay. This is a horizontal structural member connected to the combustion chamber, which is used to stiffen the combustion chamber wall. The upper arrow on the left points to a seismic stop. Part of the stop is connected to the vertical structural support member and other members are connected to the buck stay. Together they form the lateral restraint. What appears to be corrugated steel behind the upper arrows is the wall of the combustion chamber.

Figure 6.16. Spalled concrete at construction joint

Figure 6.17. Pipe support and seismic stop

Figure 6.18 shows the wall-structural member adjacent to the control room wall that failed and was replaced with a plastic dust barrier. Some tiles of the control room suspended ceiling also fell. A few feet away (Fig.6.19) shows the construction joint between the boiler support structure and the powerhouse with signs of distress and relative motion. Figure 6.20 shows the .25-inch drop in the turbine pedestal floor relative to the powerhouse floor. This is unusual because the turbine pedestal usually has a massive foundation. There were no signs of impact between these structures at this joint, which is a common occurrence, because the turbine building is more flexible than the turbine pedestal.

Tubes surrounding the low-voltage output cables from the generator were damage due to relative displacements (Fig. 6.21). It is not known if these cables, which go to the step-up transformers just outside of the powerhouse, were gas insulated or if the tubes are just used for protection.

Figure 6.18. Wall adjacent to failed control room wall

Fig. 6.19. Damage between structures

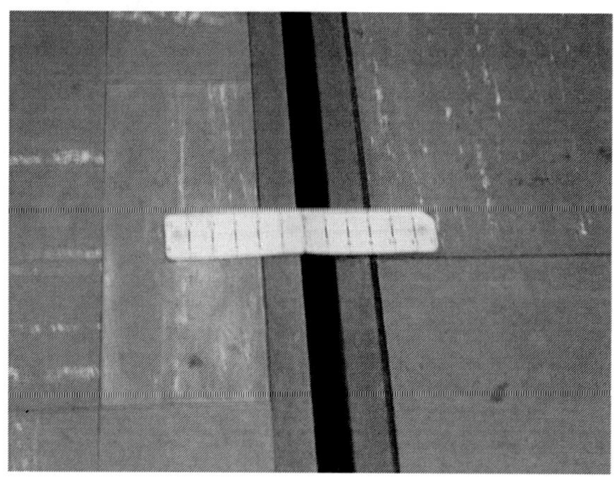

Fig. 6.20. Gap by pedestal shows no impacts

Figure 6.22 shows a new water line being installed to replace a damaged line. There were no obvious signs of ground deformation. However about 20 ft. away in the switchyard adjacent to the powerhouse, there was subsidence of about 3 in. adjacent to a bus support structure, and it was tilted. This appeared to be quite localized as no other subsidence was observed, and the tilt in the bus support structure did not cause any damage.

It was reported that a coal conveyor system in the coal storage yard was damaged. The long conveyor that caries coal from the ground to the top of the boiler support structure, which is vulnerable to damage from the lack of provision for the relative motion at the top of the boiler support structure, was not damaged.

The estimated cost for repair parts for the plant was $5 million. Figure 6.23 shows the new 370 MW plant under construction adjacent to the existing plant.

It was reported that the 60 MW Petropower cogenerating plant also had stack problems and was out of service. The Bocamina Substation associated with the old and new power plants is located adjacent to the existing power plant. There was no damage to equipment. Figures 6.24 and 6.25 show the two sections of the switchyard. The new part of the substation to be used for the new plant has not yet been energized.

Figure 6.21. Conduit for low-voltage generator output bus was offset

Figure 6.22. New water pipe replacing damaged pipe

Figure 6.23. New power plant being constructed adjacent to the existing plant

Figure 6.24. Substation servicing the existing power plant

Figure 6.25. Substation servicing the new power plant

6.4 Transmission System

The functions that must be performed by substations and their equipment are universal, and the configurations to meet these needs are relatively limited. Indeed, much of the equipment in Chile is made by the same manufacturers as the equipment in the United States, although there are some notable differences in specific types and methods of installation.

This section provides an overview of seismic qualification methods used in Chile, shows maps of the network in the earthquake-affected area, summarizes substation damage, and discusses observations at each of the substations visited. Observations made at a given substation may be applicable to others.

Seismic Qualification of Substation Equipment

Chile has established the "Security and Quality of Service Code" that created two acceptable seismic equipment qualification procedures. [2] These are the General Technical Specifications (GTS) ETG- A.0.20, which were originally developed by ENDESA but have been translated and modified by Transelec, and "IEEE Recommended Practice for Seismic Design of Substations" (IEEE 693-1997), which are the specifications used in North America. The following provides a limited comparison of the two qualification procedures and is not meant as an evaluation of either.

There are several similarities between the two procedures. The GTS has many provisions that are specified by Transelec. IEEE 693 is a recommended practice, but is not a code. If adopted by a utility, a limited number of its procedures are at the discretion of the utility, but the intent is to have uniform test or analysis procedures and acceptance criteria. Both GTS and IEEE 693 generally require shake table testing to qualify most transmission substation equipment. They both use 2 percent damping as the default value for selecting the required response spectrum (RRS), a term common to both documents. The GTS lower and upper cutoff frequencies that define the flat part of the RRS are damping dependent and are different from those in IEEE 693, but the differences are not that significant. For the 2 percent RRS the GTS lower and upper cutoff frequencies are 2.1 Hz and 8.4 Hz, compared to 1.1 Hz and 8.0 Hz in IEEE 693. While there is a large difference in the lower cutoff frequency, there is no practical difference because most equipment has natural frequencies well above these values. The acceptance criteria for brittle materials for both GTS and IEEE 693 has a factor of two, so the procedures are equivalent. GTS has an explicit requirement that the allowable stress be 2-sigma below the mean, while IEEE 693 is mute on this.

The main differences between the qualification procedures are the following: The spectral accelerations of the two RRS are 1.37 (GRS) and 1.618 (IEEE 693), so that IEEE 693 at 2 percent damping is about 18 percent more conservative in the peak of the RRS. IEEE 693 allows the test response spectra (TRS) to be 5 percent below the RRS over short intervals. Figure 6.26 shows the RRS for GTS, and Figure 6.27 shows the RRS for IEEE 693. IEEE 693 partially accounts for conductor loading by requiring a 15-pound

---

[2] Chile legislation in 2009, "Norma Tecnica de Seguridad Calidad y Servicio" established acceptable seismic qualification provisions.

weight (for 230 kV equipment) or 25-pound weight (for 500 kV equipment) be added to the terminal. The GTS requires that a force of 225 pounds (230 kV Equipment) or 393 pounds (500 kV equipment) be applied when testing. These terminal loads will be more severe than the weight added to the terminal in IEEE 693.

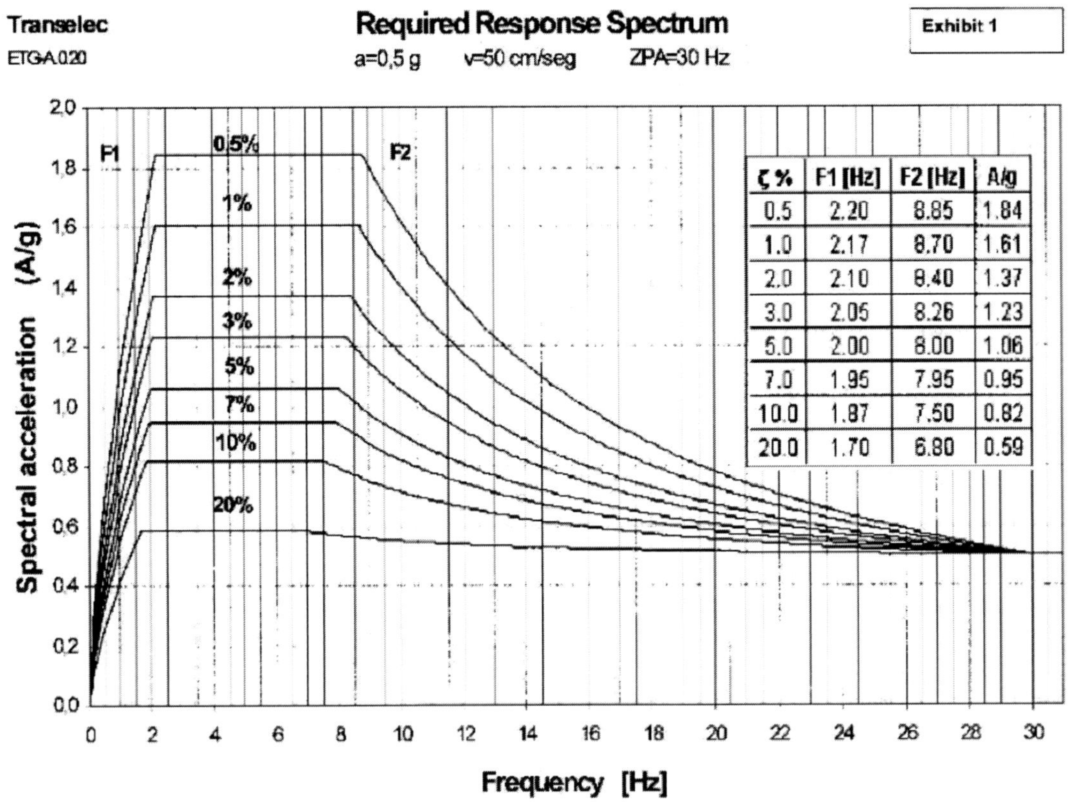

Figure 6.26. RRS used in Transelec GTS (Courtesy of Transelec)

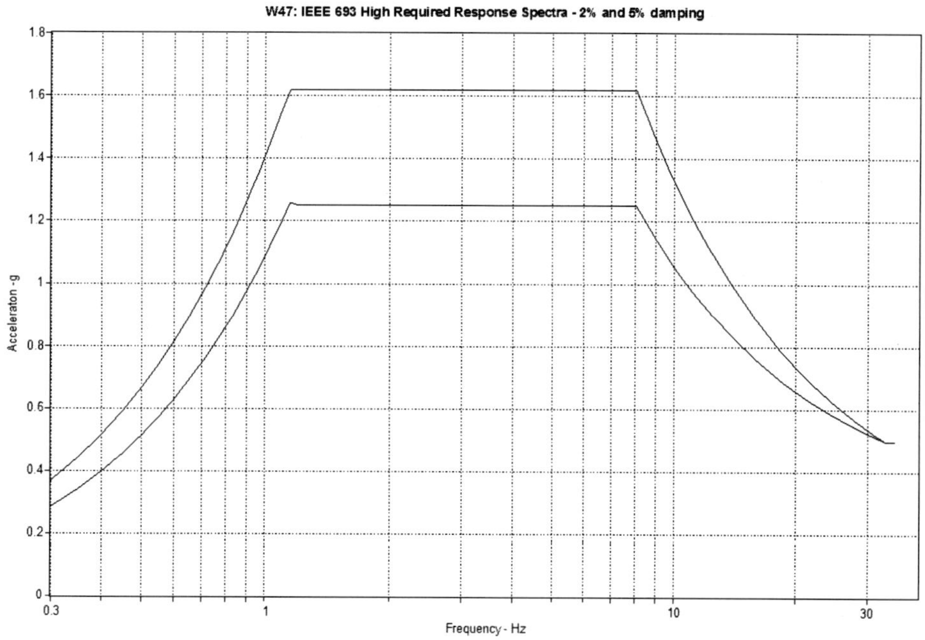

Figure 6.27. High seismic RRS used in IEEE 693 (2% and 5% damping)

Some of the differences are that the current draft of IEEE 693 has largely eliminated the sine beat test that was required for circuit breakers. It also used a sine sweep (as an option) to determine the system's natural frequencies, but at relatively low amplitude excitation. The GTS appears to use a sine sweep at relatively high excitation levels, but the comparison of these levels and sweep rate to the RRS will not be evaluated here. There is also a difference of amplification needed when equipment is tested without a support structure. GTS requires an amplification of 1.5 of the input, while IEEE 693 requires an amplification of 2.5 if information on the support structure is not known. However, for the evaluation of transformer bushings, which are not tested on transformers, in IEEE 693 the amplification is 2. The GTS scales the vertical ground motions to 60 percent of the horizontal, while IEEE 693 uses 80 percent.

Ground Motions

It would be desirable to have strong-motion recordings at the substations that were visited so that equipment performance could be related to the seismic exposure at each site. Unfortunately, none of the sites had strong-motion instruments. A recent report by the University of Chile with principal author Boroschek (2010) provides useful information on the records recorded. This reports identified the region, locality, and station name of many free field sites. It contains peak accelerations of many sites, plots of the acceleration components at some sites, and plots of the corrected acceleration, velocities, and displacements of other sites. For selected sites spectral accelerations plotted against period are also provided. However, not all information is provided for all records. In addition, Professors Boroschek and Contreras kindly re-plotted the spectra for some of these records with the 2 percent damped IEEE 693 RRS superimposed and provided site information that explained what appeared to be anomalies in some of the data. For most substations that were visited, the nearest strong-motion instrument was tens of miles from the site. As has been frequently observed in other earthquakes, records obtained from relatively close sites can be very different. For example, in Concepción we visited three substations and a power plant, and there are three nearby strong-motion sites. In the Santiago area, there are two substations south of the city and one north of the city and five strong-motion sites in the area. Table 6.2 lists sites with strong-motion records and associated peak acceleration and nearby substations. Within each group, a site with a strong-motion record is taken as a reference to the separation of nearby sites. There are two cases where substations are close together, and their separations are indicated.

It is informative to look at the accelerations of the two strong-motion sites near Concepción. These sites are just 5 km apart, but the horizontal peak accelerations at the Colegio San Pedro site are 50 percent to more than 200 percent larger than those at the Colegio Inmac. Conc. site. Professor Boroschek noted that the large low-frequency content of this record is being investigated with addition of more strong motion instruments, but it appears to be associated with the response of a valley with a predominant period where the instrument is located. Clearly making projections of ground motions at substation sites 3 to 20 times more distant from the strong-motion sites is of limited value. Very often the character of the response of equipment at a site, such as the motion of furniture or computers on desks, can provide an indication of local ground motions. This may be particularly problematic for this earthquake because of the character of the spectral acceleration exhibited in the response spectra of the records recorded in this earthquake. While the severity of the ground motion at a site is typically characterized by the peak acceleration, the response of the power equipment at the site is controlled by the spectral accelerations, and both Chilean and U.S. qualification procedures use requires response spectra that relate the peak accelerations to an envelope of spectral accelerations representing large expected responses. It appears that long-duration earthquakes generate response spectra that are very different from the RRS. For most records for which response spectra are available, in the frequency range of most substation equipment, the spectral accelerations often exceed the RRS by a factor of two. This raises a question of the utility of characterizing substation motions if these estimates are going to be off by a factor of two or more.

The authors are grateful for the reformulated response spectra for several sites provided by Professors Boroschek and Contreras of the University of Chile. Preliminary ground acceleration plots for the records for four of these sites are taken from the report cited above and are shown in Figures 6.28 to 6.31. These records are not free-field records, but were recorded at ground level. The records from a fifth site, the Concepción – San Pedro site, is discussed in detail and shown below (Figures 6.35 and 6.36). Note that the duration of strong shaking are about 60 seconds.

Table 6.2. Sites with Strong-Motion Records and Nearby Substations Visited

	Distance km	N -S Peak g	E-W Peak g	V Peak g
Sites Near Santiago				
Manipu CRMS	Reference	0.560	0.488	0.241
Puente Alto – Hospital Sotero Del Rio	107	0.265	0.268	
Penalolen Hospital Luis Tisne	107	0.295	0.263	
Polpaico Substation	113			
Paine Substation	97			
Hospital Substation (4.6 km from Paine)	99			
Sites Near Concepción				
Concepción – Colegio San Pedro	Reference	0.618	0.650	0.58
Concepción – Colegio Inmac. Conc	5.2	0.401L	0.286T	0.398
Concepción Substation	8.0			
San Vicente Substation	10.9			
Coronel Substation	18.7			
Bocamina PP & SS (2 km from Coronel)	20.6			
Site Near Angol				
Angol	Reference	0.928	0.681	0.281
Charrua Substation	85			
Site Near Talca				
Talca	Reference	0.477 L	0.424 T	0.244
Ancoa Substation	39			

Of greater interest are the response spectra. In addition to the records in Table 6.2, the spectra from the Puente Alto-Hospital Sotero del Rio located near Santiago are shown in Figures 6.32 to 6.36. Of the 10 spectra shown here, two per site, and the two associated with the Concepción – Colegio San Pedro (Fig. 6.28), all but two exceed the IEEE 693 RRS that is superimposed on the spectral plots. Of particular note are the spectra in Figure 6.36, which have very high spectral accelerations at low frequency.

Figure 6.37 shows all of the spectra on a single plot so that a better overall assessment of how the spectra compare to the RRS can be made. Two features stand out in this figure: Many of the spectra exceed the RRS in the frequency range from 4 Hz to 8 Hz, and one site has very high low frequency spectra at about 0.6 Hz.

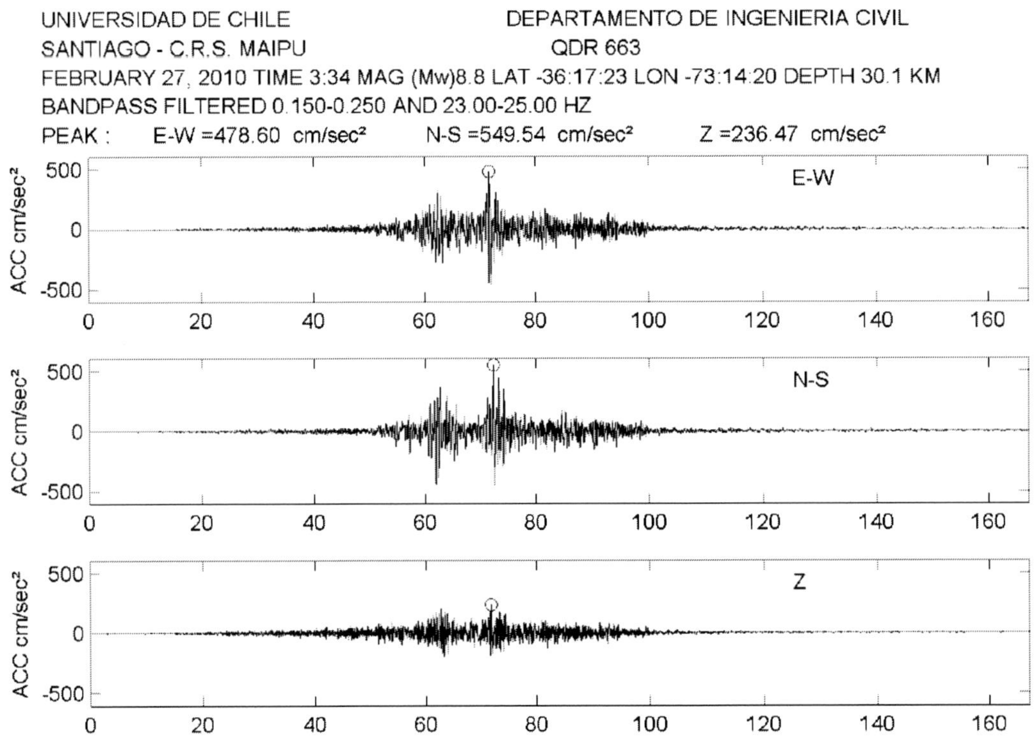

Figure 6.28. Santiago – Maipú CRS Acceleration (Courtesy of R. Boroschek and V. Contreras, University of Chile)

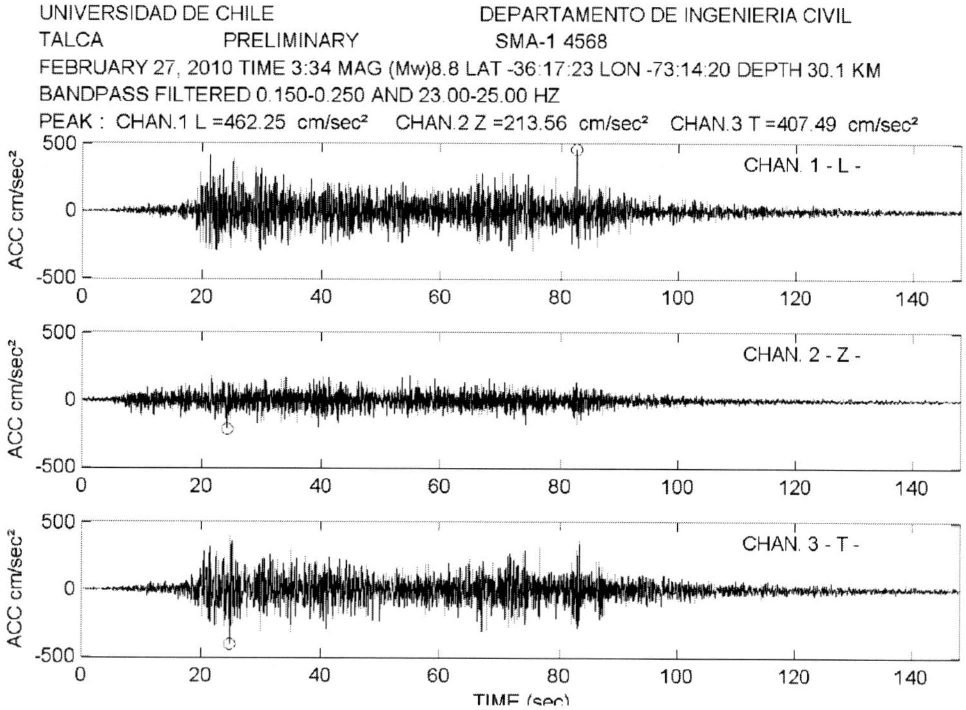

Figure 6.29. Talca accelerations
(Courtesy of R. Boroschek and V. Contreras, University of Chile)

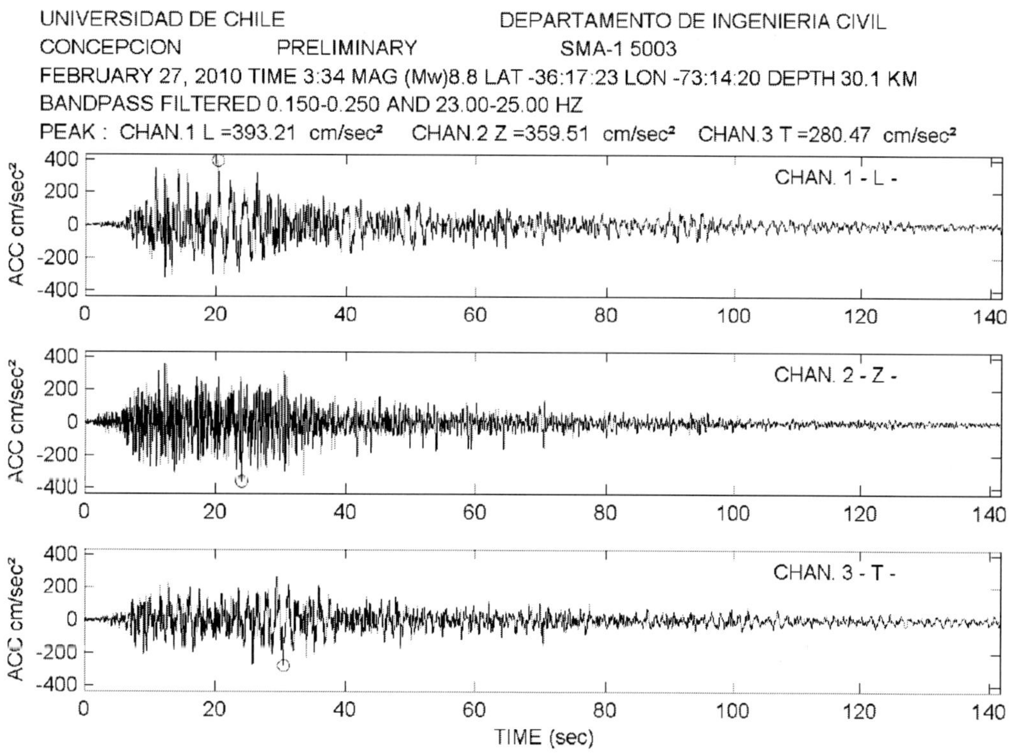

Figure 6.30. Concepción – Colegion Inmac. Conc.
(Courtesy of R. Boroschek and V. Contreras, University of Chile)

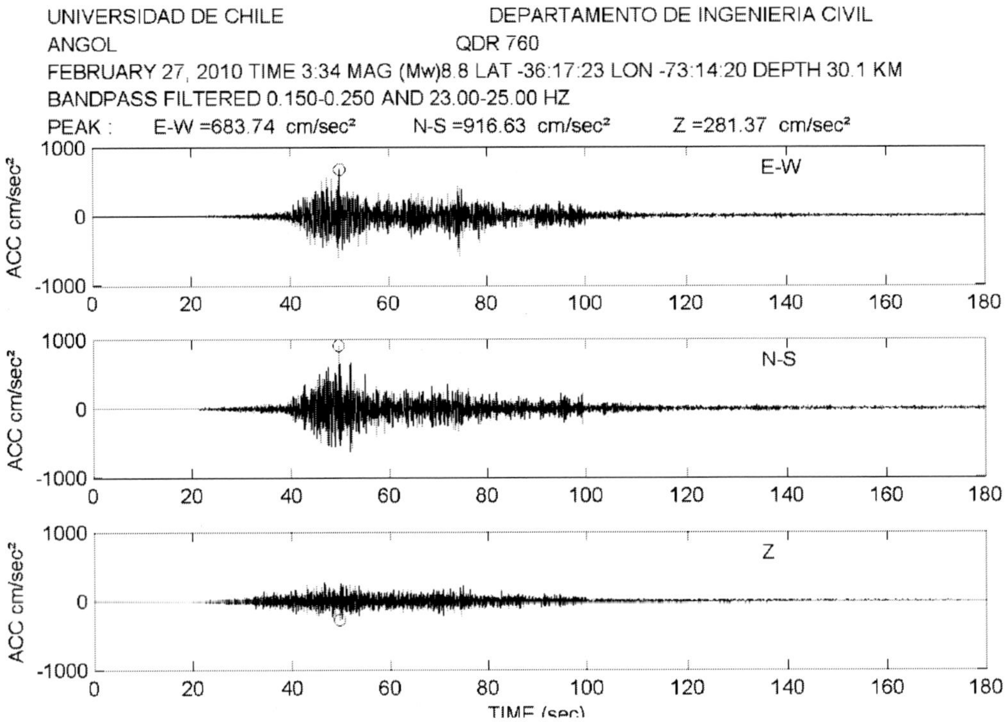

Figure 6.31. Angol Accelerations
(Courtesy of R. Boroschek and V. Contreras, University of Chile)

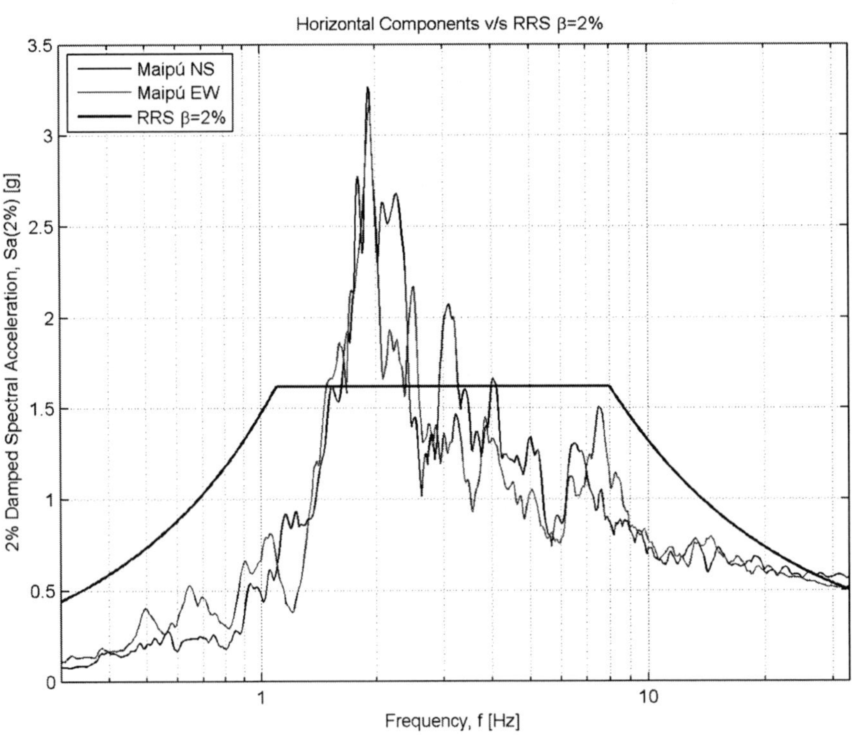

Figure 6.32. Maipú response spectra
Courtesy of R. Boroschek and V. Contreras, University of Chile)

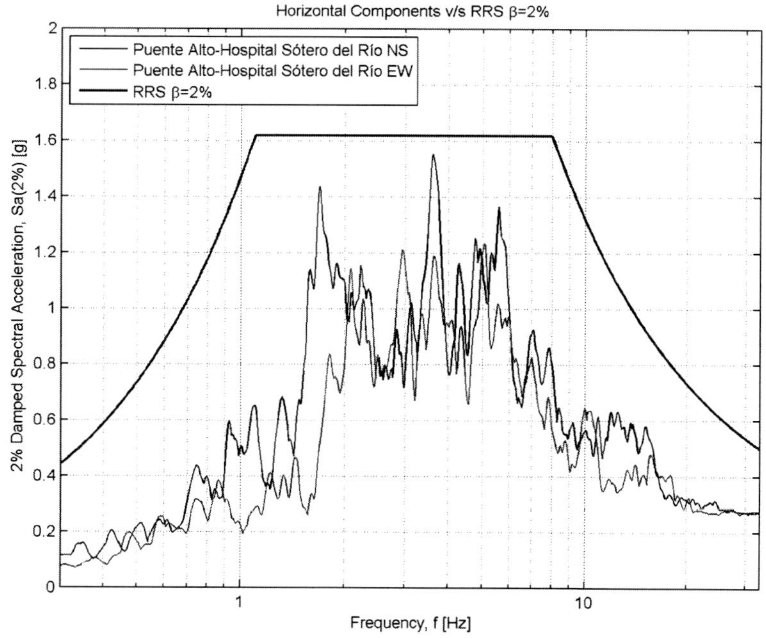

Figure 6.33 Puente Alto response spectra (Courtesy of R. Boroschek and V. Contreras, University of Chile)

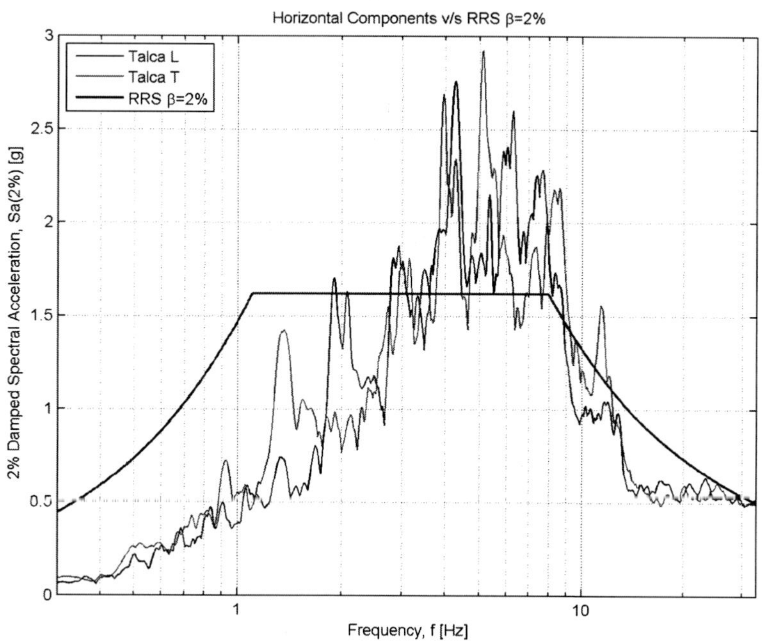

Figure 6.34 Talca (Courtesy of R. Boroschek and V. Contreras, University of Chile)

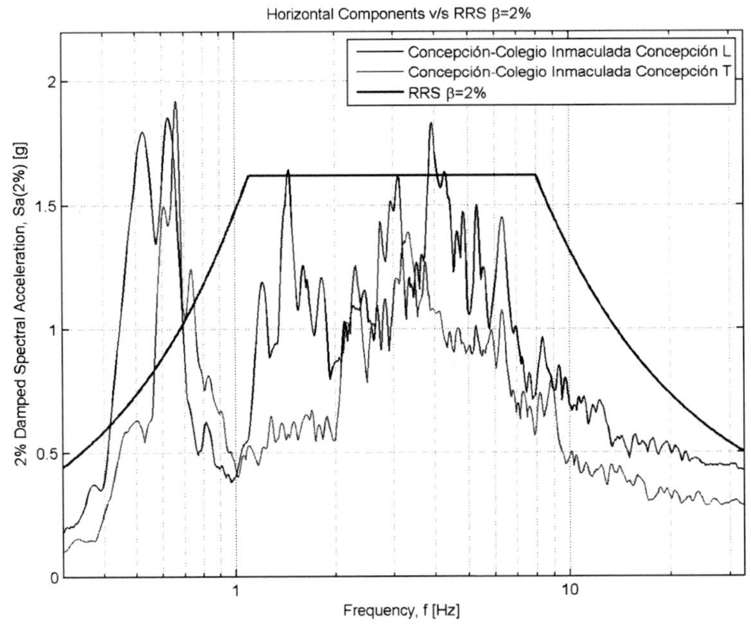

Figure 6.35 Concepcion – Colegio Inmac. Conc. (Courtesy of R. Boroschek and V. Contreras, University of Chile)

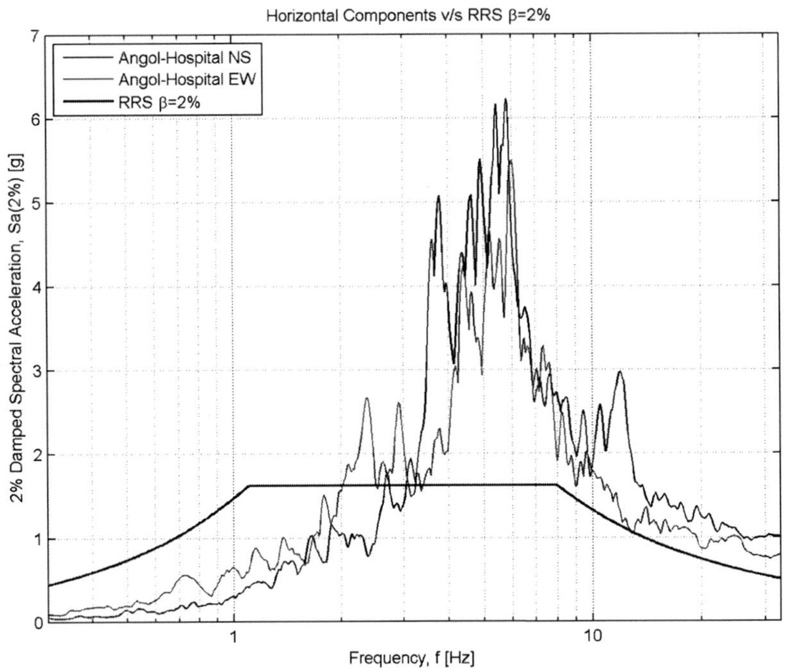

Figure 6.36 Angol – Hospital (Courtesy of R. Boroschek and V. Contreras, University of Chile)

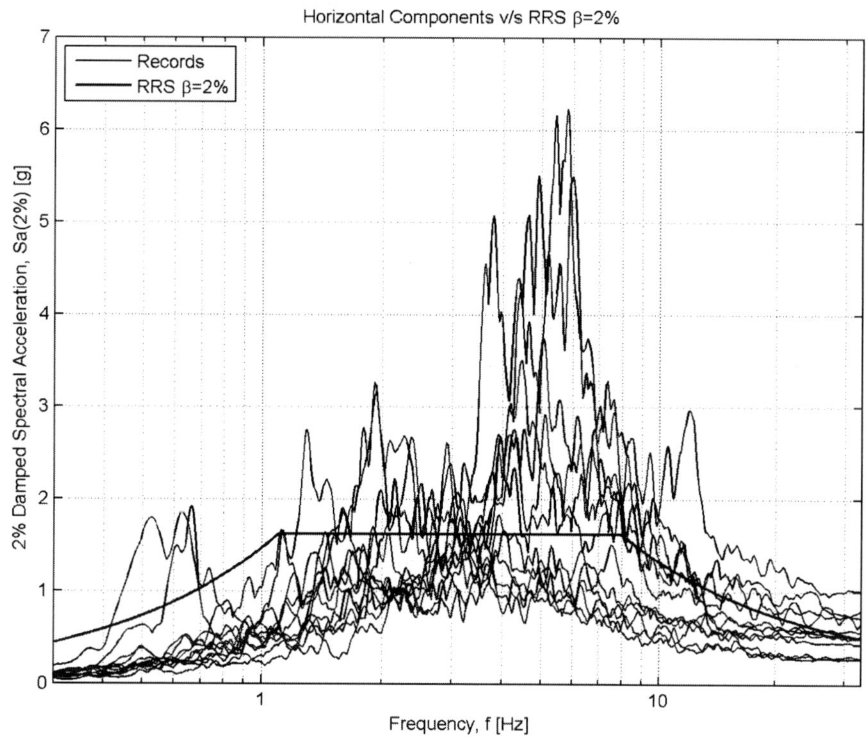

Figure 6.37 Original spectra plotted with IEEE 693 RRS (Courtesy of R. Boroschek and V. Contreras, University of Chile)

It is useful to look at these same spectra scaled so that the peak acceleration of the individual records is 0.5g (Fig. 6.38). Several observations can be made from this plot, some of which have implications to the classification of substation sites in the United States for determining if the site should have equipment qualified to moderate or high RRS. While the RRS was not designed to envelop the spectra from all earthquakes, the following provides a useful summary of observations about these response spectra. In the case of IEEE 693, as can be determined at this time, the RRS have not been formulated by establishing a level based on the standard deviation of a collection of records, such as one standard deviation above the mean. The level has been set to envelop a large percentage of the earthquake spectra used in the formulation. Most records were from short duration events with strong shaking less than 15 seconds duration.

- The long duration of shaking appears to have generated relatively high spectral accelerations in the interval from 2 to 8 Hz. This has recently also been observed in the Baja, California, earthquake. The typical method in IEEE 693 for determining if the equipment at a substation site should be qualified to the moderate or high level, is assessing the peak acceleration at the site. The relatively high spectra accelerations for a given peak acceleration in this event suggests that if the site can be exposed to long-duration earthquakes, a moderate site with peak ground acceleration large than 1.3 g should be considered for upgrade to a high seismic site for the purpose of qualifying equipment. While a formal study of what constitutes long duration strong shaking has not been done, pending additional evaluation, it is suggested that earthquakes with durations of strong shaking 25 seconds or longer be considered as long earthquakes.

- At frequencies above about 8 Hz, the RRS does provide an envelope for most of the records.
- With the exception of the records from one site, Concepcion – Colegio Inmac. Conc., spectral accelerations below about 1 Hz were about half the value suggested by the RRS. It was anticipated that this large magnitude earthquake would have generated larger low-frequency spectral acceleration.
- The large low-frequency spectral accelerations at the Concepción – Colegio Inmac. Conc. site suggest that if base isolation or frequency shifting methods are to be used to protect substation equipment, a site-specific evaluation of the site should be required.

Some members of the investigation team visited the Concepcion – San Pedro site, and it is informative to look at the setting where the strong-motion records were obtained. This record was recorded at the Colegio San Pedro Concepción (Concepción High School), located just west-southwest of Concepción. The record has not been corrected, so it should be interpreted with caution. Three team members visited the site, so some information about the site is known.

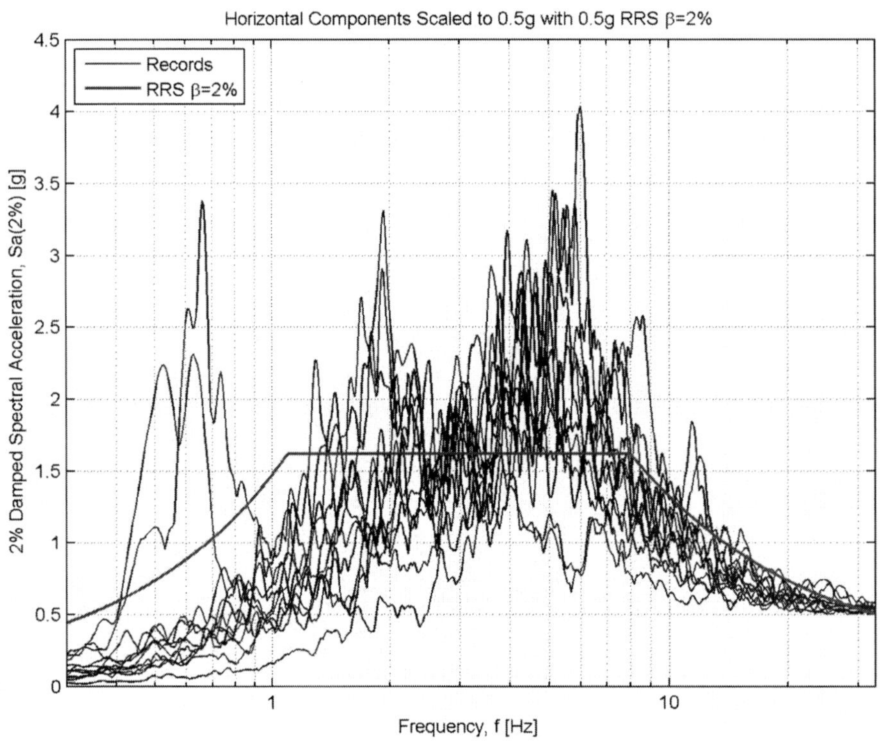

Figure 6.38. Spectra of records normalized to 0.5 g peak acceleration (Courtesy of R. Boroschek and V. Contreras, University of Chile)

The site is classified as a rock site and rock outcroppings can be seen in the area, but there was a thin soil layer over the rock near the instrument vault. The instrument was located in a concrete vault on a concrete pedestal (Fig. 6.39). An overview of the vault and its relation to two adjacent buildings is shown in Figure 6. 42. The buildings at the facility form an L-shape. The long leg of the L appears to consist of three independent buildings. At the top of the L is a three-story building founded on drilled piers. The two other buildings that form the lower part of the L are one-story structures. The instrument vault is located in the inside corner of the L. The buildings that form this part of the L are only one-story high and appear to be founded on spread footings. The arrow in Figure 6.40 identifies a 3.8-cm (1 1/2-in.) separation between the buildings that form the corner; there does not appear to be any impact damage at the crack edges. Figure 6.41 shows cracks in the soil that appear to mark the edge of the spread footing. There are no such cracks on the other side of the building. There are additional cracks behind a vault cover in the foreground and other cracks in the soil were observed in the area. To the left several feet out of view in this photograph there is a slight downhill slope. The instrument vault is in the background.

It is clear that this is not a free field instrument location because of its proximity to two structures. The extent to which the adjacent structures have affected the recorded ground motion, however, is unclear. The cracking in the soil by the edge of the footing, other cracks in the soil, and the movement of the structure that forms the bottom leg of the L suggest soil movements in the area. If the pedestal supporting the instrument is anchored into the underlying rock, the effect of the adjacent buildings and soil movement may not have a large effect on the record, but this was probably put in for ease of maintenance. Thus, there are questions about the soils and possible interactions with nearby buildings.

Noting the uncertainties of the instrument setting, several observations can be made. The accelerations for the horizontal and vertical motions are shown in Figure 6.42, with a peak acceleration of 0.65 g for one of the horizontal records. The most notable feature about the time history is the duration of strong shaking, which is about 70 seconds.

Figures 6.43 and 6.44 show the response spectra for the horizontal and vertical records that have been scaled to have a peak acceleration of 0.5 g with an over-plot of the 0.5 g Requires Response Spectra (RRS) from IEEE 693. The horizontal response spectra exceed the RRS by about 70 percent in the frequency range around 5 Hz, a common resonant frequency of substation equipment. The RRS was not designed to envelop all earthquake spectra, so having a response spectrum exceeding the RRS is not uncommon. Cursory reviews of other spectra from other records from this earthquake indicate that a large peak in the spectra in this frequency range is characteristic. The vertical response spectrum also shows high frequency content, which would be consistent with a rock site. If there was pronounced interaction with the adjacent buildings, a response with a narrow peak in the spectrum corresponding to the frequency of the building would be expected, and this is not present.

Figure 6.39. Instrument on pedestal in concrete vault

Figure 6.40. Location of instrument that recorded the largest record

Figure 6.41. Cracks may be from soil consolidation of low-slope landslide

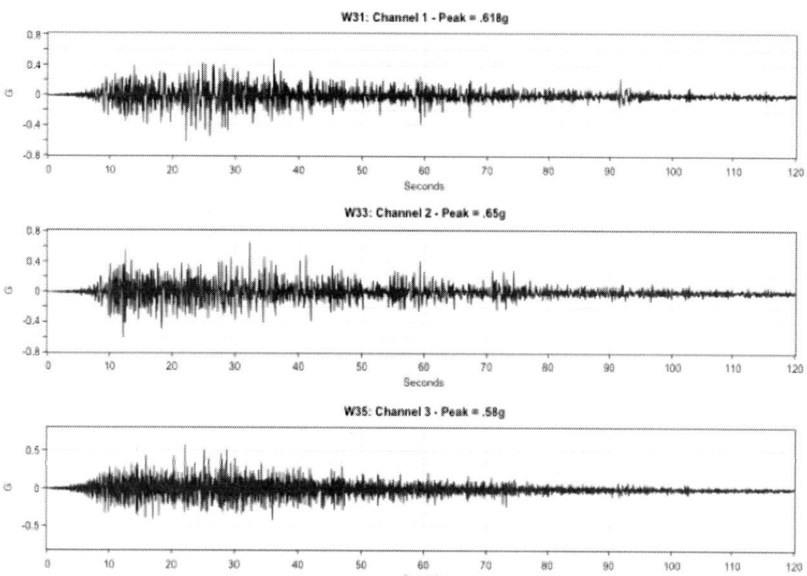

Figure 6.42. Strong-motion records with long duration

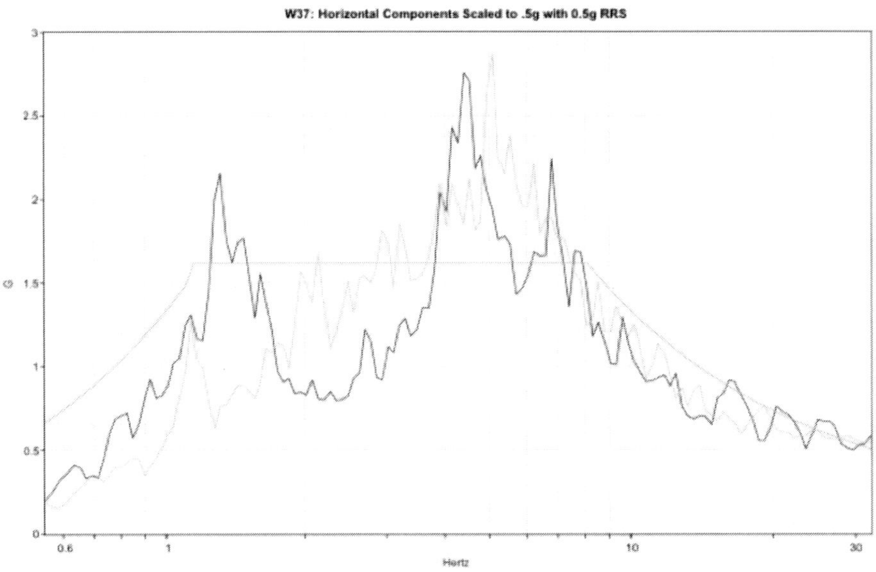

Figure 6.43. Horizontal test response spectra of 0.5g scaled records with IEEE 693 RRS

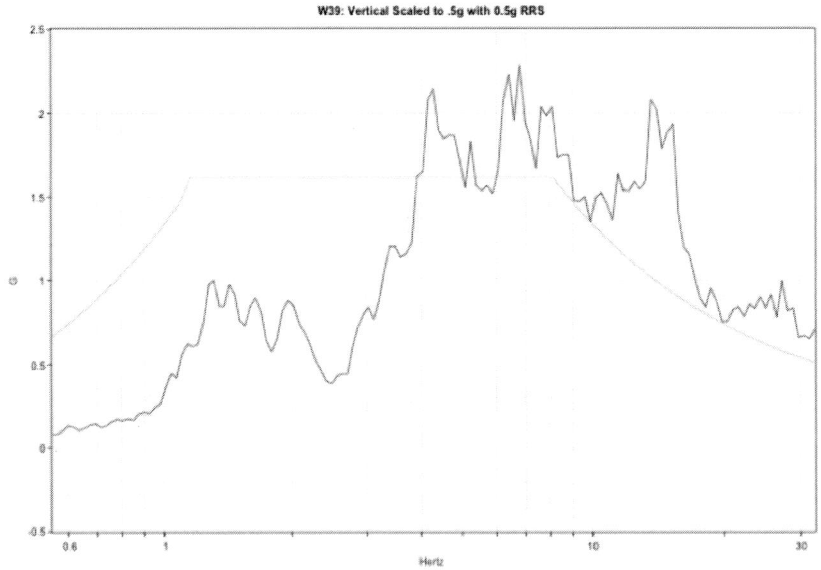

Figure 6.44. Vertical test response spectra of 0.5g scaled records with IEEE 693 RRS

Power Network Maps

Figure 6.45 shows the region north and south of Santiago that was impacted by the earthquake. This figure represents less than 20 percent of the length of Chile. Figure 6.46 shows the region around Concepción affected by the earthquake. The arrows on each figure indicate power facilities the team visited. The team was only able to view three other facilities from outside the gate; these are not indicated by arrows. These maps, which were created in 2009, not only show the Transelec transmission network in solid lines but also transmission circuits of other companies with dashed lines. The two dark lines that run from the Polpaico substation in the north to Charrua substation in the south indicate the 500 kV lines in the system. It is impossible to distinguish in a black and white rendering the green 220 kV and red 154 kV lines, but most of the solid lines between the stations that were visited (see arrows) are 220 kV circuits. The thin dark lines are 66 kV circuits. As an example of the circuits serving a substation, Charrua is a very large substation that has two 500 kV transmission line circuits, 13 220 kV circuits, five 154 kV circuits, and five 66 kV circuits.

Transelec owns two sections of the Chile high voltage network—the Northern Interconnected system (SING) and the Central Interconnected System (SIC). The Northern Interconnected System has 959 km of transmission line and four substations. The Central Interconnected System has 7,289 km of transmission lines and 46 substations. The system operates primarily at 500 kV, 220 kV, and 154 kV. The estimated loss to Transelec due to equipment damage at the time of the visit was $7 million.

Detailed Descriptions of Transmission Substation Damage

Members of the investigation team obtained information about damage to substation equipment during visits to substations by word of mouth or from a picture taken from a car as it passed a substation. Information about the character of the damage is limited. It is assumed that bushing problems are referring to slipping of bushings that resulted in an oil leak, if not otherwise specified. Some published reports show damage to substation equipment. This may have already been included in our tally, so we did not add it in again. It is fair to assume that additional damage, particularly to distribution system circuit breakers and possibly disconnect switches, is not included in this report.

Polpaico Substation

Polpaico Substation was the first substation visited by some members of the investigation team. It is located northwest of Santiago and services two 500 kV circuits and nine 220 kV circuits. During the visit the team only had time to walk through the 500 kV switchyard, although there also was a large 220 kV switchyard. It was reported that one of twenty-one 220 kV candlestick circuit breakers failed, and three 220kV CB of the same type that were under construction also failed. Damage in the 500 kV switchyard included nine of ten 500 kV transformer or reactor bushings and three surge arresters.

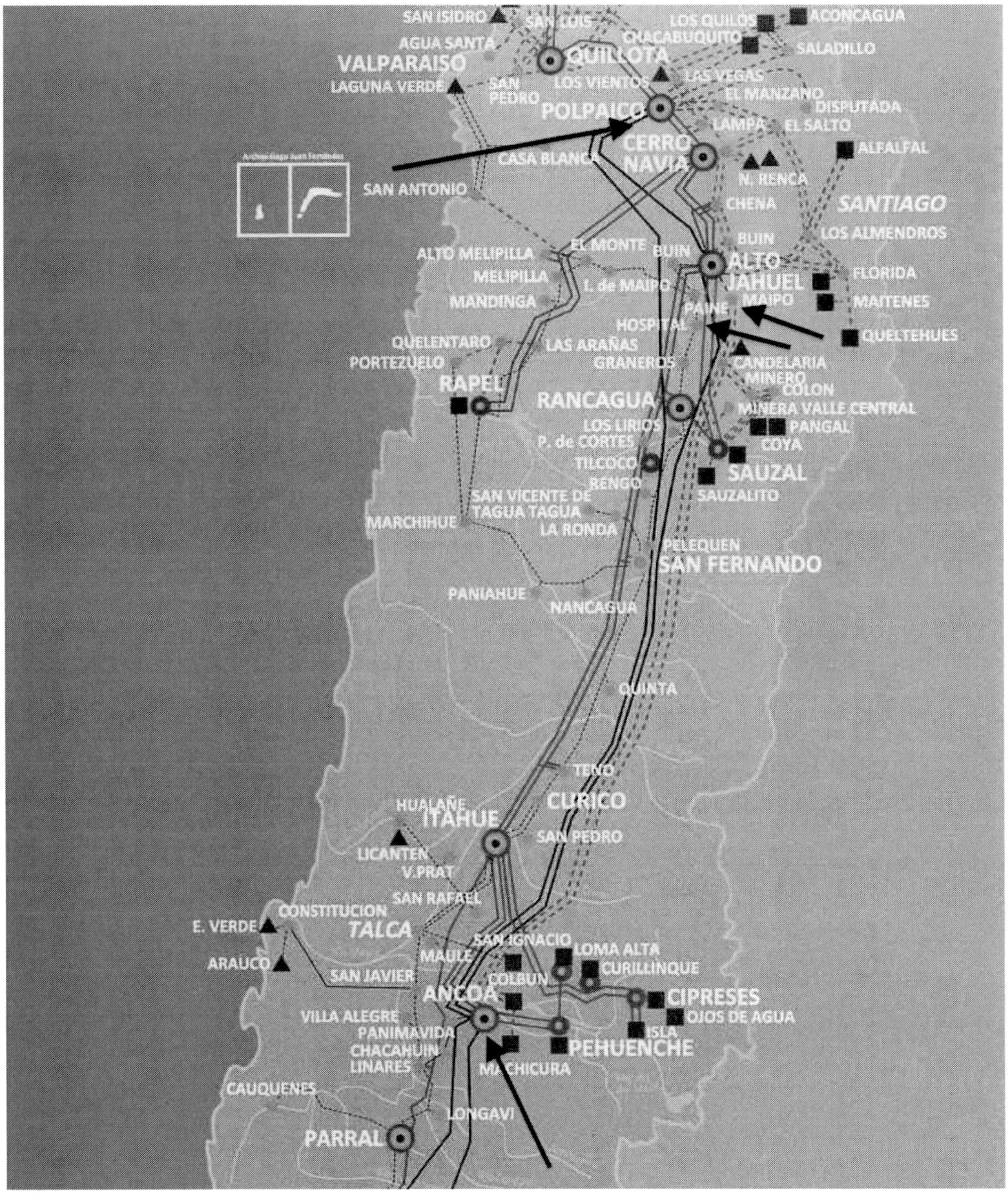

Figure 6.45. Power grid around Santiago and to the south in 2009. (Courtesy of Transelec)

Summary of Transmission Substation Damage

Table 6.3 summarizes damage at transmission substations observed and described to the team during its visit. Below the table is a list of abbreviations used in the table. Figure 6.47 shows a desk in the control room with two flat screen displays. During the earthquake these tipped over or slid off of the desk, suggesting significant ground acceleration. The small size of the control room was striking for the size of the site; however, this substation is controlled from the regional control center, and the local control room is used only for emergency operations and substation maintenance.

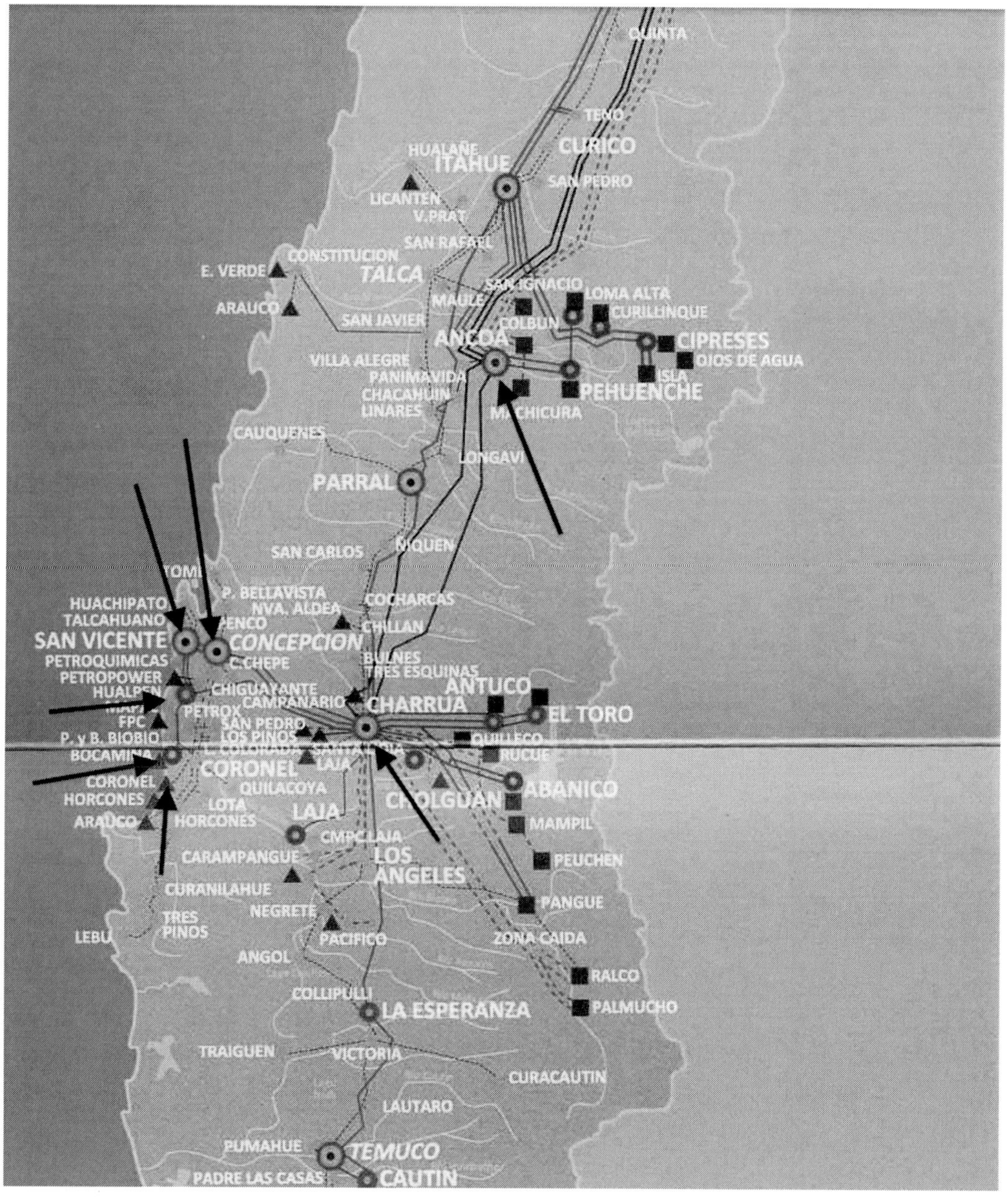

Figure 6.46. Power grid around Concepción and to the north in 2009 (Courtesy of Transelec)

One of the more interesting types of damage at this site was to the relatively new porcelain 500 kV bushings used on three 3-phase reactors and one single-phase transformer (spare). These bushings were spring-loaded, center clamped but also had external clamped porcelain. External clamped porcelain bushings at this voltage are relatively rare in the western United States. One of the new bushings on a reactor that failed is shown in Figure 6.48. This photograph is from the A-Phase or C-Phase of a reactor bushing because the bushing is installed at an angle. The flange casting failed on all of the damaged bushings. The 500 kV bushings on the three in-service transformers were undamaged.

Table 6.3. Damage and Observations at Transmission Substations

Substation Voltages	Operating Voltage			Miscellaneous, Comments
	500kV	220 kV	154 kV	
Polpaico 500, 220	* 9 /10 center & external clamped porcelain bushing	4/13 CCB failed-flange, 1 in service and 3 under construction		* Flanges cracked
Charrua 500, 220, 66	2 pantograph DS damaged 30 PI attached under dead-end structure failed 7 of 14 CT terminals damaged	PT terminal failed 4 SA on firewall 18 monopolar CCB base plates cracked TR bushing sheds *TR bushing slipped	3 CTs of 31 failed 1 CC failed at mid-level TR sheared anchor bolts	4 Weld crack near Buchholz relays Spring supports to HVAC *Epoxy to stop leak
Ancoa 500, 220, 66	6/30 pantograph DS PI under dead-end structure			Extensive use of damping devices
Substation near Ancoa 220		Problem with dead-tank CB		
Concepción 220, 154, 66		2 DS TR radiator oil leak		CR ceiling
San Vicente 154 kV, 66 kV			1 pantograph disconnect switch failure reported	Wall failure Lateral cracks soil crakes Pipe cracks near Buchholz relay on TR 66 kV 15 failures of rigid bus supports on rack

Transformer (TR), Surge Arrester (SA), Circuit Breaker (CB), Candlestick Circuit Breaker (CCB), Disconnect Switch (DS), Potential Transformer (PT), Current Transformer (CT), Live-Tank (LT), Control Room (CR), Post Insulator (PI), Coupling Capacitor (CC)

In the 1985 Manzanillo, Mexico, earthquake, seven of the 29, 400 kV clamped porcelain bushings had failures in the porcelain at the base of the bushing near where they were clamped. The peak acceleration measured in the switchyard was 0.38 g. There were also six 230 kV center clamped, uncemented bushings at the site; all slipped at the flange-porcelain interface and leaked, so bushing problems are not restricted to any specific design.

Figure 6.49 shows a reactor in which the A-Phase and C-Phase failed bushings had been replaced. Several things can be observed in this figure. While it is difficult to estimate the slack in the conductors shown here, it would be interesting to get actual slack measurements. The terminal connections failed for three of the 500 kV surge arresters due to high flexibility. Subsequent communications indicated that the damage was deformation to the terminal and grading ring due to conductor loading. One of the damaged surge arresters was associated with the undamaged bushing (Fig. 6.49). The first connection to all of these bushings is the 500 kV surge arrester. Note that the turrets supporting the bushing are quite tall. As shown in Figure 6.50, however, each is anchored to the tank at the flange, and there is a substantial brace between turrets. Arrows point to the braces to the cover and between turrets. While it is difficult to estimate material thicknesses from these pictures, it appears that the cover plate on the turret is 1.25 cm

Figure 6.47. The desktop equipment was completely disturbed by earthquake

Figure 6.48. Fractured flange on porcelain clamped bushing (Courtesy of Transelec)

Figure 6.49. A 500 kV reactor with the damaged A-Phase and C-Phase bushings replaced

(1/2 in.) to 2.54 cm (1 in.) thick. The gravity load due to the tilt on the A and C phases places additional loads on the base of the bushing and the flange casting. The bracing of the turrets and the relatively small diameter of the plate supporting the bushing suggest that the as-installed frequency of the bushings will be relatively high compared to the bushing mounted on the spare transformer. All tilted bushings on the reactors failed. Two out of three vertically mounted bushings failed.

Figure 6.50. Reactor turrets are braced to cover and to each other

Figure 6.51 shows an older transformer (name plate data: ANSI C59.12-1980). This transformer has a high-voltage bushing turret with a large footprint, moderate-size manifold mounted radiators, a fire suppression system, and what appear to be two conservators (a large and small unit). Transmission transformers are usually installed with firewalls. The bushing appears to be a center-clamped, uncemented bushing like those used in the United States. The turret's large footprint and its proximity to three of the tank walls will tend to increase the as-installed frequency of the bushing system.

Figure 6.52 shows the bolted anchorage on the reactors. Figure 6.53 shows the anchorage on the transformer shown in Figure 6.51, which used an existing wheel assembly to bolt the transformer to the foundation. All transmission system transformers at all sites visited used anchored rods that provided robust anchorage. None showed signs of distress for the moderate ground motions experienced. Many of the utilities in the western United States use welded anchorage to plates imbedded in the foundation slab.

Figure 6.54 shows the upper structural support of the radiator shown in Figure 6.51. These radiators lack diagonal braces, which are recommended in the United States, to provide lateral stiffness. However, they have a structural connection supporting the radiator assembly to the transformer tank at the top and bottom. In the United States, the pipe that connects the radiator to the top of the tank is typically used as the mechanical support for the radiator. This often results in oil leaks at the flange connection to the tank.

Figure 6.51. Older transformer at Polpaico Substation

Figure 6.52. Typical transformer anchorage using anchor rods

Figure 6.53. Transformer anchorage using wheel mount

Figure 6.54. Structural system holding the top of radiator to transformer tank

Figure 6.55. Spark-gap rods were provided with the transformer bushing

Figure 6.55 shows spark-gap hardware on the high-voltage bushing. These were provided with the bushings, and the gap was set to 125 percent of the protection level because the surge arrestor and the circuit breaker provide protection.

Figures 6.56 and 6.57 show through-bolt surge arrester standoffs. In this design the insulators are subjected to compressive loads, and the through-bolt resists overturning moments. The arrows indicate the insulators above and below the support structure. Figure 6.58 shows a standoff on a surge arrester commonly used in the Unites States, but rarely seen during our investigations in Chile. In this standoff, tensile loads, due to overturning moment, are carried by the grouted connections of the caps to the porcelain and by the porcelain, which has a low tensile strength.

Figure 6.59 shows a heavier live-tank circuit breaker that is mounted on the pendulum system shown in Figures 6.10 and 6.11. This circuit breaker has two interrupting heads, each with two interrupting chambers, for each phase rather than the one interrupting head shown in Figure 6.10. This unit has a capacitor in addition to a resister associated with each interrupting chamber. None of these units was damaged.

Figure 6.60 shows a wave trap installation that is typical of those observed at all sites.

Fig. 6.56. Top of standoff

Fig. 6.57. Bottom of Standoff

Fig. 6.58. Non-through bolt

Figure 6.59. Heavier, dual-head circuit breaker similar to that shown in Figure 6.10

Figure 6.60. Wave trap suspended directly below its support structure

The wave trap is suspended just below the dead-end structure by two insulator strings to reduce motion in the structure's plane. This method is suggested in the ASCE Manual of Practice 98 in the United States. When a surge arrester is suspended to the level of the equipment, it tends to vibrate back and forth and damage the equipment it is connected to even when it has generous slack to adjacent equipment and a lower restraint. Figure 6.61 shows damage after an earthquake in the United States, caused in part by the swinging of the wave trap suspended at the equipment level.

Figure 6.62 shows an overview of the 220 kV switchyard. The team did not have time to conduct a detailed walkthrough; however, it was reported that one candlestick circuit breaker failed in this switchyard and three 220 kV CB were under construction. In a subsequent discussion, it was noted that the cast flange failed rather than the porcelain. Note that this figure shows 48 circuit breaker phases. Of these, 21 are the candlestick type; none have any type of base isolation or added damping devices.

Figure 6.63 shows a microwave tower used by the substation perched on a ridge just outside of the substation. As frequently happens in the United States, the dish was misaligned by the earthquake and had to be realigned to reestablish communications.

Figure 6.61. Swinging wave trap damaged adjacent equipment (United States)

Figure 6.62. Overview of part of the 220 kV switchyard

Figure 6.63. A microwave dish was misaligned by the earthquake

Charrua Substation

The Charrua Substation is located east of Concepción in the southern part of the earthquake damage zone and services two 500 kV circuits, thirteen 220 kV circuits, and four 66 kV circuits. The water table at the site is 1 to 1.5 m below the surface. In the 154 kV switchyard, three current transformers and a potential transformer failed. In the 220 kV switchyard, two surge arresters mounted on firewalls failed and a terminal on a potential transformer failed. Additionally, seven of 14 candlestick circuit breakers failed. Welds on piping connections adjacent to four Buchholz relays broke on transformers. In the 500 kV switchyard, two surge arresters mounted on top of a firewall adjacent to transformers failed, and one damaged a 500 kV transformer bushing. At least seven station post insulators mounted to the underside of dead-end structures failed, as did two pantograph disconnect switches.

Three of six computer terminals on a desk tipped over during the earthquake (Fig. 6.64), and a refrigerator in an adjacent room slid across the floor, indicating that the ground motion was probably in the moderate range. Figure 6.65 shows marks left on the wall as the fire extinguishers rocked in the earthquake. The water table at the site is quite high, so some transformers appear to be on platforms above grade (Fig.6.66). These are actually tanks to accommodate oil from the transformer, should there be a leak. Because of the high water table, some control cables normally placed in conduit at grade were placed above grade (Fig.6.67).

Figure 6.68 shows a disconnect switch design not observed in the western United States. Figure 6.69 shows a center-break disconnect switch and support structure similar to those used in the United States. This older vintage switch uses cap and pin insulators rather than a station post insulator. Figure 6.70 shows the connection used for the control rods to open and close the switch. These are adjusted in the field and held with U-bolts. To make a non-friction connection in the United States, a piercing bolt is used after the switch is aligned. In Chile they achieve the same effect by drilling a hole through the coupling and the control rod and make a bolted connection. Figure 6.71 shows a coupling of two rods with friction connections.

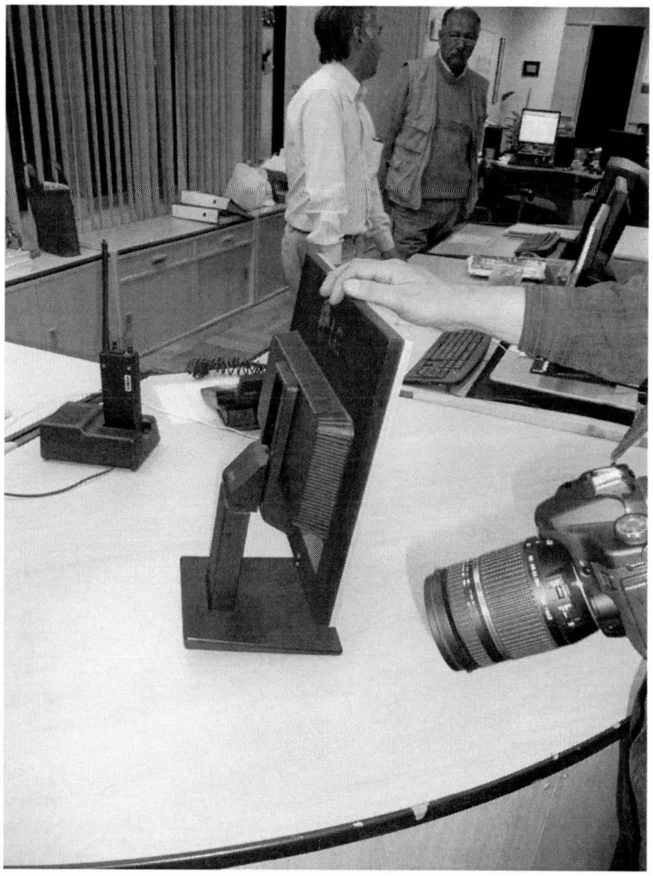

Figure 6.64. Terminals tipped over in earthquake

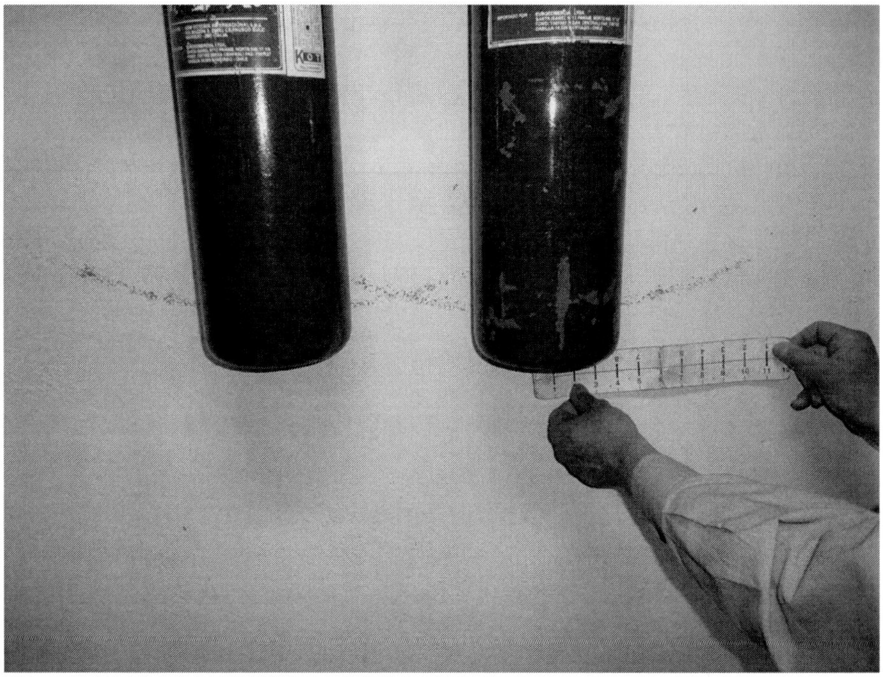

Figure 6.65. Marks on the wall from the swinging of fire extinguishers

Figure 6.66. Some transformers placed above grade because of the high water table

Figure 6.67. Above-grade control cable conduit

Figure 6.68. A disconnect switch design that is uncommon in United States

Figure 6.69. Center-break disconnect switch similar to those found in the United States

Fig. 6.70. Positive connection with bolt

Fig. 6.71. Coupling using friction

Figure 6.72 shows a 154 kV capacitor coupled voltage transformer that failed at the base of the upper section. The vertical conductor drop indicates that conductor slack was not an issue in the connection to the adjacent equipment. Conductor dynamics in the long vertical conductor drop could have contributed to the failure. This is an unusual failure because the upper and lower sections are the same diameter, so that if the failure was caused by a cantilever load at the terminal, the porcelain connection failed at half of the load experienced by the lower section. This exhibits the large variability in the strength of the porcelain, which may be a quality control issue in its manufacture or it may be that an impact-type load (due to conductor dynamics) and a higher modal response contributed to the failure. Figures 6.73 and 6.74 show 220 kV potential transformers, one with a broken connection. Figure 6.75 shows an overview of the two potential transformers and suggests that slack may be an issue.

In the 154 kV switchyard three current transformers at one position failed. It was reported that one had a terminal failure, and the other two have internal damage.

Figure 6.76 shows a 154 k, 66 kV (75 MVA) transformer taken during the team investigation. The team member in the shadow to the right of the transformer provides perspective on the height of the transformer, which is rather tall. Figure 6.77 shows a transformer anchorage failure. Because of the similarity of the radiator bracing it is assumed to be identical or the same transformer as shown in Figure 6.76; however, there may have been some modifications. In the picture taken at the time of the investigation, there was no channel deformation or corrosion, and there appears to be two channels rather than one.

Figure 6.72. Capacitor coupled voltage transformer failed at mid-height

Fig. 6.73. Good PT connection

Fig. 6.74. Failed PT connections

Fig. 6.75. Slack questionable

Figure 6.76. Overview of a 154 kV transformer similar to that in Figure 6.60

Figure 6.77. Anchorage damaged in the earthquake (Courtesy of Transelec)

Several observations can be made from Figure 6.77. The left of the two larger bolts in the foreground, which are bolted to an embedment in the support pad, is slightly bent. This indicates that after the transformer anchorage failed, it slid and hit the bolt. It is interesting that only the left bolt appears to be bent, but the locking nut is positioned to come into contact with the channel's flange before the bolt on the right. The left of the two smaller bolts had lifted more than 1 inch (2.5 cm) from the plate it was attached to, stripping the threads on the bolt or in the plate, so that the transformer must have lifted off its pad. The transformer moved to the left and away from the camera exposing the bolthole in the plate. Note that the smaller bolt on the right has been fractured and that the fracture surface is about one-third the diameter of the bolt above the plate. It is not clear if this is a guillotine shear facture or a brittle fracture. The offset of the channel suggests that bolts at more than one anchor point failed. In the figure there appears to be only one channel, so that the only bolts holding this corner of the transformer are the two bolts shown. The photograph taken after restoration appears to have two channels, so there are probably four bolts holding the channels to the foundation plate.

Historically, earthquake investigations have shown that bolted transformer anchorages have not performed as well as welded anchorage, that is, the transformer base plate is welded to embedments. However, these configurations are different from those used here. The poor performing anchorages had anchor rods cast in place in the foundation pad, and the holes in the transformer base plate for the anchor rods were oversized to accommodate errors in rod placement. While these anchors were generally conservatively designed, it was assumed that because of the clearance in the holes it would not engage all of the anchor rods at the same time when the transformer slipped so that initially the first rod to come in contact would have to support the entire load. Thus, there would be incremental failure as successive rods engaged. The recommended practice is to install a thick, close-fitting plate washer, and after it is in place, weld it to the transformer base plate. In this way all anchor bolts or rods would engage to share the load and restrain the transformer. In the current situation, the anchor bolts did not have excessive clearance, so load sharing would have been achieved.

Figure 6.78 shows the base of one of two 220 kV (1050 kV BIL) live-tank candlestick circuit breakers that had cracked base plate flanges. In walking through the switchyard about five other units may have been damaged but were kept in service by temporarily placing two L-shaped backup plates in opposite corners below the support plate (Fig. 6.79). Arrow A on the left shows the backup plate. Arrow B on the left shows a bolt that also holds the insulator flange to the support plate, but has been extended through the support and backup plates. The Arrows C on the right point to the bolt that goes through the cast aluminum circuit breaker base plate and the steel mounting plate on the support structure. It is not known if the corner of the cast aluminum circuit breaker base plate was undamaged or cracked. The steel plate is the support structure was undamaged.

There were several problems related to transformers. Figure 6.80 shows an overview of the first transformer in a bank of four 220 kV to 161 kV units. First, note that the conservator and its support structure extend well above the top of the transformer (arrow on the left). Second, note the surge arrester on top of the firewall (arrow on the right). It is connected to what is believed to be a 220 kV bushing in the foreground (arrow in center). Figure 6.81 shows a surge arrestor on the ground that was replaced by the unit shown in Figure 6.80. The arrow in the foreground of Figure 6.81 points to the broken terminal tab that was connected to the bushing. The arrow in the back shows the base of the surge arrester. The surge arrester base connection used through bolts similar to those shown in Figure 6.56. It is believed that the base was removed from the top of the firewall after the surge arrester failed. Figure 6.82 shows the base of the surge arrester that pulled out of its lower flange, an unusual failure mode. Figure 6.83 is the surge arrester from the next transformer with failed feet cast in the lower flange. The surge arrester that replaced this unit used standoffs like those shown in Figure 6.58. Figure 6.84 shows the base of the bushing connected to the failed surge arrestor on the first transformer.

This appears to be a center clamped, non-cemented bushing; however, there is a slight offset between the flange and the porcelain (arrow on left) and material has been added to the interface between the flange and the porcelain (arrow on right). It is speculated that this is an epoxy coating that has been added to stop a slow oil leak that was draining the bushing. This technique was used after the 1994 Northridge earthquake where some twenty-five 230 kV bushings of this type developed leaks, and there was a need to keep some transformers in service. A slow leak in the bushing will eventually cause the bushing to fail, possibly catastrophically, and could also damage the transformer.

Figure 6.78. Corner of base plate of 220 kV candlestick circuit breaker cracked off

Figure 6.79. Field repair to damaged base plates of candlestick circuit breakers

The conservator support frame on the 220 kV to 161 kV transformer is shown in Figure 6.85. Figure 6.86 shows the bolted connections and an enlargement of a slipped connection. Each of the four transformers in the bank developed a crack in the flange weld in the assembly adjacent to the Buchholz relay. This is assumed to be a result of the shifting of the conservator support frame. Figure 6.87 shows the Buchholz relay; the adjacent pipe flange that has been replaced is shown with an arrow. The location of the failed weld is unknown.

The problems that have been attributed to slotted connections in support structures include the introduction of impact loads when the bolt hits the end of the slot. This can introduce high frequencies that can be detrimental to porcelain supported by the structure. If one of the bolts hits the end of the slotted connection before the other bolts, it can experience a high shear load that could shear the bolt. In this way the bolts could shear in sequence and cause total failure of the support. In a visit to Chile by one of the team after the 1985 earthquake, one of the conservators sheared some of its support bolts, which rotated about the lower support and came to rest leaning against the adjacent firewall. Slotted connections are very helpful in assembling support structures fabricated off site because they can accommodate slight imperfections in the foundation. In the United States it is recommended that after assembly slotted connections be tack welded to prevent slipping.

Figure 6.80. Arrows show the tall conservator, surge arrester and 220 kV bushing

Figure 6.81. Failed surge arrester and failed terminal tab

Figure 6.82. Base of surge arrester pulled out of its lower flange

Figure 6.83. Legs on base of surge arrester failed

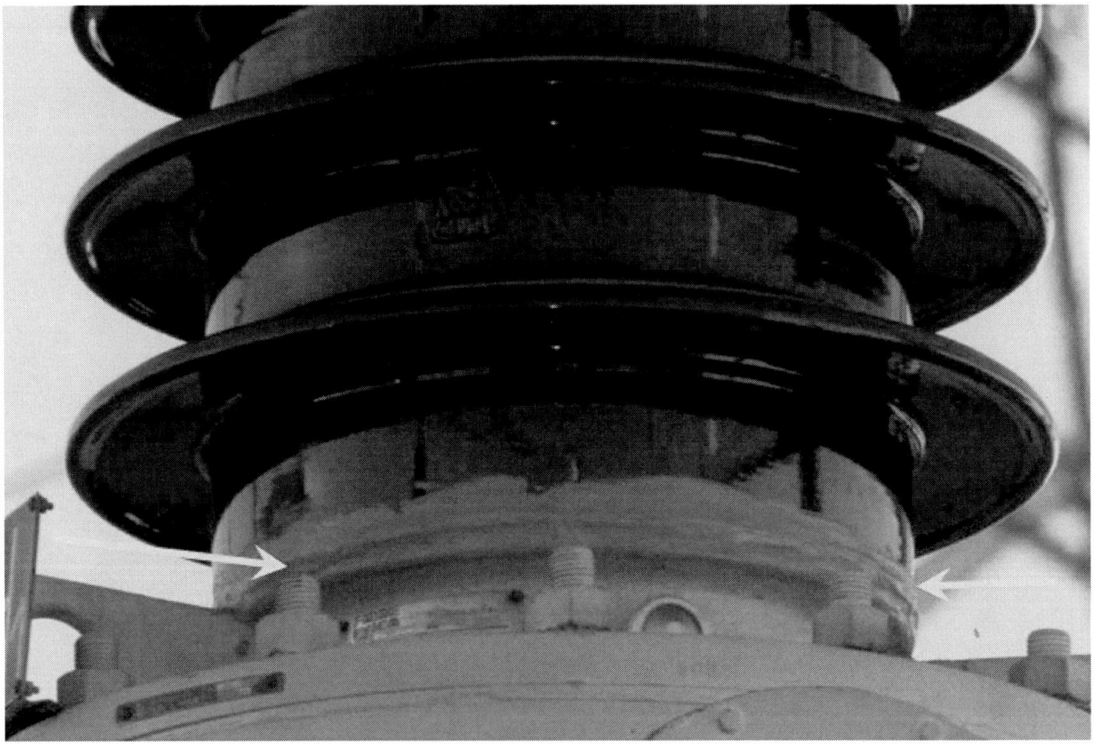

Figure 6.84. Bushing with offset and patch material to stop a leak

Figure 6.85. Frame supporting conservator is bolted to the tank

Figure 6.86. Bolts securing conservator with insert showing slipping

Figure 6.87. Replaced damaged pipe flange connection

Figure 6.88 shows one of two 500 kV reactors associated with the 500 kV transformer bank. The units were undamaged. Note the large conductor sag between the bushing and the surge arrester assuring adequate slack in the connection. Figure 6.89 shows the large and highly braced turret that supports one of the cemented bushings. Figure 6.90 shows the seven transformers separated by firewalls that make up the two transformer banks with a spare. Note that there are surge arresters on top of the firewalls. Two of these failed and one severely damaged sheds on the 500 kV bushing it provided overvoltage protection for. Figure 6.91 shows one of the damage surge arrestors, which had a porcelain failure at the lower flange. Figure 6.92 shows that the surge arrester had ample slack. The failure could be due to the amplification of the out-of-plane response of the firewall or loads associated with cable dynamics. It would be interesting to measure the natural frequencies of each of the elements—surge arrestor, firewall, and cable. There are plans to relocate the surge arrester off the firewall, but the lack of space around the transformer and the need to make conductor connections are problematic. Figure 6.93 shows the damage to a 220 kV transformer bushing caused by the falling surge arrester.

In Chile none of the spare transformers were provided with pre-installed bus connections that would facilitate replacing a transformer in the bank. Some utilities in the United States have bus work installed and a pre-made conductor with end fittings to quickly replace any transformer without requiring the transformer to be moved or to fabricate bus work to connect the transformer into the circuit.

Figure 6.88. One of two 500 kV undamaged reactors associated with the transformers

Figure 6.89. A large turret supporting the A and C phases on reactor

Figure 6.90. Seven transformers separated by firewalls that support surge arresters

Figure 6.91. One of two surge arresters installed on top of firewall that failed

Figure 6.92. Surge arresters have generous slack

Figure 6.93. One falling surge arrester damaged the 220 kV transformer bushing

Figure 6.94. Glass suspension insulator replaced failed post insulator

There were several locations in the 500 kV switchyard where post insulators supported below dead-end structures failed. Figure 6.94 shows a suspension insulator that replaced the failed post insulator. While the lower end of the post insulator would not be expected to move much, the string insulator will swing and the slack provided, which is probably the same as that used for the post insulator, may not be adequate. The use of a composite suspension insulator, which would have one-third the weight as a porcelain or glass insulator string, would reduce the load on the instrumentation transformer. Figure 6.95 shows a dead-end structure in which the B-phase post insulator in the cross-under has failed. It may be interesting to calculate the load and compare it to the capacity of the insulator. The BIL of the pantograph disconnect switches in the switchyard is 1800 kV, so this value will be used to characterize the post insulator. Post insulators come in various strengths; from scaling the photographs, the insulator is estimated to have a 2,500-pound (1134 kg) cantilever strength. This unit has a weight of 908 pounds (412 kg) and is tapered. From Figure 6.76 the length of the span on each side of the insulator is estimated as being twice the insulator length, which is 152 inches (3861 mm). There are three conductors, and the weight per foot of each conductor is estimated to be 2.0 pounds. The total weight of the conductors is estimated as 304 pounds. The center of gravity of the conductor will be approximately at the center of the insulator, so the equivalent weight at the end of the insulator is 152 pounds (69 kg). The equivalent weight of the insulator concentrated at the end of the insulator is estimated as 400 pounds. Thus, the total equivalent weight of the conductor and post at the end is 552 pounds (250 kg). As the cantilever strength is 2,500 pounds (1134 kg), fairly large spectral amplifications would be required to fail the post insulator. Even if a lower strength post is used (cantilever strength of 635 kg [1,400 lbs.] and weight of 288 kg [636 lbs.]), a large amplification would be needed to fail the post (assuming a spectral amplification of 3.25 for 1 g ground motion and IEEE 693 RRS). This suggests that conductor dynamics may be involved, as dynamics can induce impact type loads. Figure 6.96 shows a frame in which two insulators have failed. These insulators are connected to grounding switches, have generous slack, and have less conductor weight than in the previous figure. Figure 6.97 shows the post insulators on the ground where they fell. Figure 6.98 shows a typical fracture at the base of a post insulator (i.e., a guillotine-type fracture across the insulator at the sand ring). The sand ring is a roughened part of the stub of the insulator that fits inside a pocket in the flange. Portland cement and sand is typically used to make the connection between the flange and the stub of the insulator. This is strictly a mechanical connection; there is no chemical bonding between the grout and the porcelain or the flange. This is why roughness is added to the porcelain stub, as can be seen in Figure 6.98. Secondary failures that result from the insulator striking the ground are also used in characterizing failure type.

We learned through discussions with one major insulator manufacturer that in their testing to determine cantilever strength, in which a uniform pull test to failure is used to generate the failure, the crack in the porcelain usually starts at the sand ring and extends diagonally up though the body of the porcelain, and guillotine-type failures are rare. In lab tests of post insulators conducted in a research project by a utility at a university lab, a small number of post insulators were subjected to pseudo-dynamic cyclic loading. About half the samples had guillotine-type fails and half had fractures through the insulator body.

Figure 6.95. Failure that is not related to slack, but may be related conductor dynamics

Figure 6.96. Drop with adequate slack in which two post insulators failed

Figure 6.97. Post insulators on the ground after falling from dead-end structure

Figure 6.98. Typical guillotine-type failure observed of post insulator

Figure 6.99 shows the upper part of the B phase of a pantograph disconnect switch missing and the connection being made with a jumper. Figure 6.100 shows a connection being fractured. It is not known if the other connection was removed or if the bolts making the connection failed. Figure 6.101 shows the arms on the ground with the arrow on the left with the fractured part still connected and on the right with a clean pipe. Figure 6.102 shows the fractured connection close up. While an exact count of pantograph disconnect switches is not known, a review of team photographs indicates that there were at least twenty-eight 500 kV disconnect switches. All pantograph switch failures were related to the character of the upper connection. All failures occurred in switches where the upper connection was to an inverted post insulator rather than from a flexible bus. Figure 6.103 shows a HVAC system in which the spring supports shifted.

Ancoa Substation

Ancoa Substation is located about two-thirds of the way from Santiago to Concepción and services four 500 kV circuits, four 220 kV circuits, and one 66 kV circuit. Of the substations visited, the Ancoa substation made the most extensive used of damping devices on instrument transformers (Figures 6.12 and 6.13), candlestick circuit breakers (Figure 6.6), live-tank circuit breakers (T head) in the 500 kV switchyard, and some instrumentation transformers and surge arresters in the 220 kV switchyard (Figures 6.9 and 6.88). There were 45 semi-pantograph disconnect switches and 30 pantograph disconnect switches. Seven pantograph switches poles failed. One 500 kV post insulator supported under a dead-end structure failed, damaging sheds on a potential transformer. Figure 6.104 shows the terminals in the control room, none of which tipped over.

A hydroelectric facility, owned by Colbun generating company, is adjacent to the substation and has its own small switchyard. The team did not visit the facility, but it was reported that there were problems with a dead-tank circuit breaker. Figure 6.105 shows a 220 kV dead-tank SF6 circuit breaker, which was reported to have had some problems, in the switchyard adjacent to the Ancoa substation. With one unusual exception in Japan, there is no record of this type of circuit breaker being damage in an earthquake. Historical data suggests that the problem was not with the circuit breaker but with a connection to it.

Figures 6.106 and 6.107 show the bases of a 220 kV current transformer and surge arrester, which used the clamped porcelain method of connecting the porcelain to the base flange rather than a cemented connection, a method not typically used in the United States. Note that a damping device supports the surge arrester. The current transformer is also supported by damping devices, which are not shown in the photograph (Fig. 6.91).

Figure 6.99. The B-phase pantograph disconnect switch failed

Figure 6.100. Top of pantograph switch with arms missing

Figure 6.101. Pantograph arms on the ground

Figure 6.102. Broken coupling that connects arm to switch body

Figure 6.103. Support for HVAC system shifted

Figure 6.104. Terminals that were undisturbed by the earthquake

Figure 6.105. An SF6 circuit breaker was reported to have problems in the earthquake

Figure 6.106. Base of 220 kV current transformer using clamped porcelain

Figure 6.107. Base of 220 kV surge arrestor using clamped porcelain

Figure 6.108 shows a large series capacitor rack in the 500 kV switchyard. The rack is supported on porcelain post insulators with universal joints at their ends so that the insulators are just subjected to compression. Diagonal composite suspension insulators resist lateral loads. Figure 6.109 shows a detail of the base support.

Figure 6.110 shows a potential transformer with damaged sheds (see arrow) caused when a post insulator supported under a dead-end structure failed and fell. The surface of the sheds had to be coated to prevent the unglazed porcelain from absorbing moisture. In calculations following Figure 6.74 for a similar configuration, the cause of the failure is not clear because there is adequate slack. Again, the failure may be related to conductor dynamics or low strength of the inverted station insulator.

Figure 6.111 shows a live-tank circuit breaker with a proprietary base isolation system. A system similar has recently been qualified to IEEE 693 standards by means of shake-table testing. The unit shown was also qualified by shake-table testing using Transelec specifications. The unit has a relatively low frequency so that special testing methods had to be developed. Figure 6.112 shows the four dampers used to isolate the system. This system appears to work in the same manner as the unit tested in the United States; however, there is an external difference in that the unit tested was sealed at the top, as required in IEEE 693. As noted above, this site makes extensive use of devices qualified by shake-table testing to enhance damping. Calculations indicate that they should withstand short-circuit loads and wind loads of 100 km/hr.

Figure 6.108. Large undamaged series capacitor bank

Figure 6.113 shows three poles of the pantograph disconnect switches in which one phase has been damaged. Note that a jumper has been added to bypass the switch. In this case both castings that held the pantograph arms have fractured. This is similar to the situation shown in Figure 6.100, where only one casting failed. The upper connection of the disconnect switch was attached to a post insulator, as were all other failed pantograph switches. While much of the equipment in this substation used damping devices, this was not the case with the pantograph disconnect switches. Note that all connections have adequate slack. This failure suggests that relative motion between the dead-end structure and the ground may have imposed high loads on the switch, but only one of the three phases failed despite all phases being subjected to the same loading.

Concepción Substation

The Concepción Substation is relatively close to the coast and the town of Concepción. It is serviced by one 220 kV circuit, five 154 kV bays, and nine 66 kV bays. While Transelec owns the station, it contains 154/66 transformers and 66 kV bus bars owned by CGE Transmission. Some of the Transelect busses were reenergized about 10 hours after the earthquake. The remaining Transelect and CGE Transmission busses were energized about 31 hours after the earthquake, but the load was very low. The normal pre-earthquake load was about 500 MW, but the operations of two large industrial customers, with a total normal load of 180 MW, were severely curtailed. A chemical plant was only drawing about 3 percent of normal load, and a steel mill was only drawing 2.5 percent of normal load. At the time of the visit, peak load was just 300 MW, an indication of the impact of the earthquake on industry. Two 220 kV air blast circuit breakers failed.

Figure 6.109. Base connection of capacitor bank

Figure 6.110. Potential transformer damaged by falling post insulator

Figure 6.114 shows the computer terminals in the control room. The LCD displays tipped over on the desk and tall furniture tipped over. In an adjacent building, the LCD displays tipped over on the desk or fell to the floor. Shelves moved as much as 3 ft. (1m) and tall furniture tipped over. At many facilities it was common to find damage to T–bar suspended ceilings. The control room did have damage to a plaster ceiling (Fig.6.115), so that protective scaffolding was placed over the control panel to prevent damage (Fig.6.116). This site had an engine-generator (Fig.6.117). The tank above the unit was thought to be the day tank, but it was the main fuel tank, so long-term operation was not planned. The battery located behind the unit (Fig.6.118), was one of the only unanchored batteries observed in all of the sites visited.

Figure 6.119 shows five out of nine failed interrupter heads on a 220 kV air-blast, live-tank circuit breaker. The bases of these circuit breaker columns were fitted with Bellville washers to enhance damping (Fig. 6.14). Figure 6.120 shows one of the two failed porcelain posts observed during the site visits. These did not have a guillotine failure at the sand ring.

Figure 6.121 shows one of the only leaking transformer radiators observed in the investigation. Surprisingly this radiator had good diagonal braces limiting lateral movement of the radiator, a detail that usually prevents leaks. Figure 6.122 shows the transformer bushing; it could not be determined if it was cemented or not.

Figure 6.111. Live-tank circuit breaker on base isolation system

Figure 6.112. One of four dampers is identified

Figure 6.113. Only one of three phases of the pantograph disconnect switch failed

Figure 6.114. LCD terminals on the desk tipped over in the earthquake

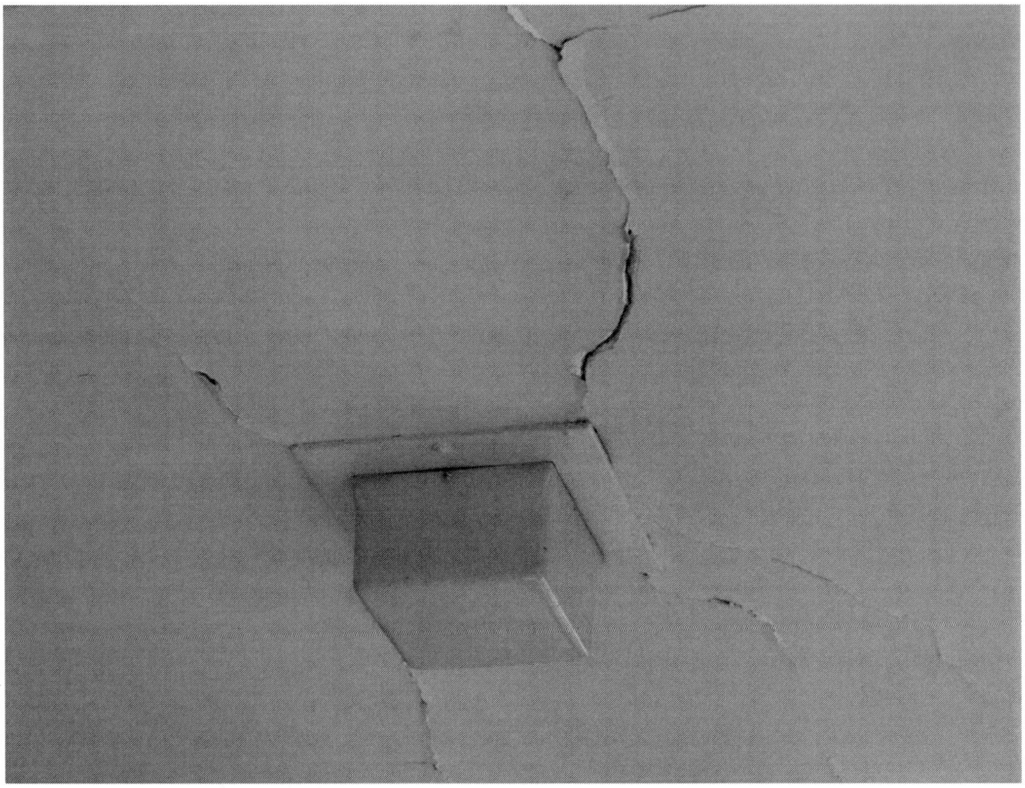

Figure 6.115. There was extensive cracking in the plaster ceiling

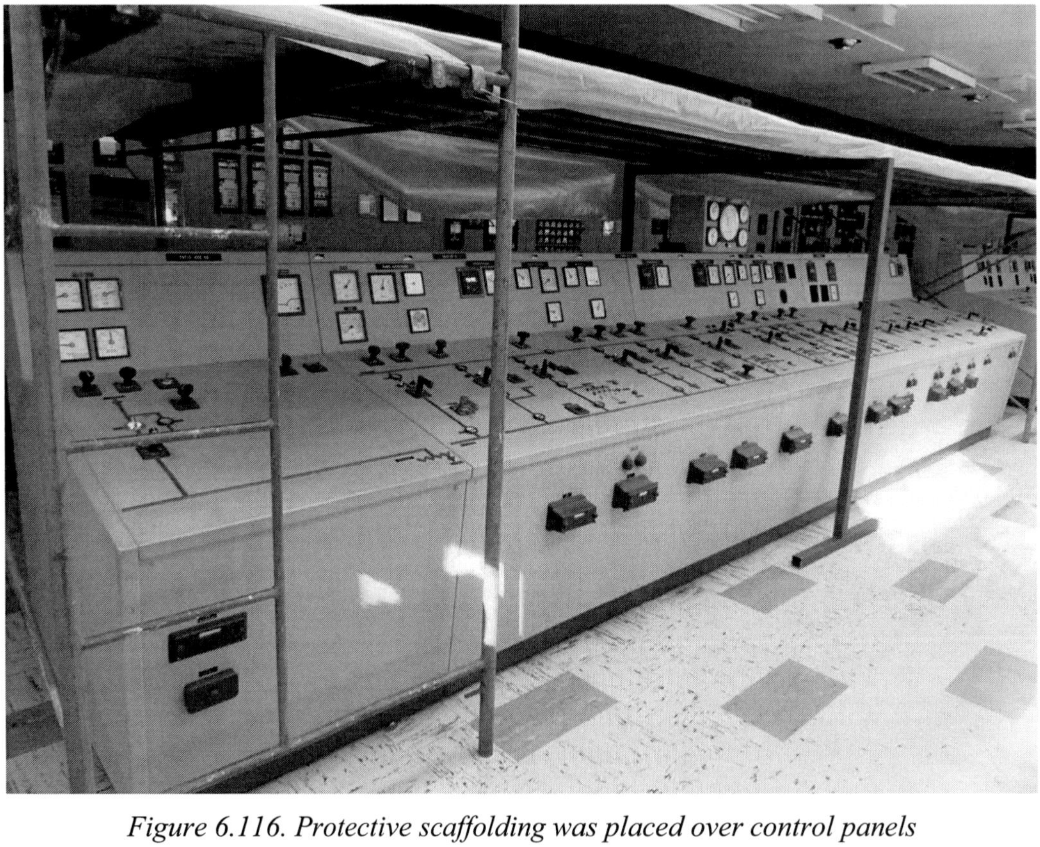

Figure 6.116. Protective scaffolding was placed over control panels

Figure 6.117. Main fuel tank supported over engine-generator

Figure 6.118. The only unanchored battery rack observed on the investigation

Figure 6.119. Five of nine heads on a 220 kV air-blast, live-tank circuit breaker failed

Figure 6.120. Damaged posts show guillotine and fracture into body of post

Figure 6.121. A radiator with stiff lateral braces still developed an oil leak

Figure 6.122. Transformer bushing that may be cemented or uncemented

Figure 6.123 shows an interesting detail, called a pigtail, for providing slack in conductors between equipment. It not only provides slack, if there is significant relative motion between the ends of the connection, the pigtail will deform providing a record of the deformation. This circuit was a 154 kV, 2000 amp service, so it may not be usable for heavier conductor. This site had the only bulk oil dead-tank circuit breaker observed during site visits. While they are being phased out in the United States, many are still in service.

San Vicente Substation

The San Vicente Substation is near the Concepción Substation but closer the coast, and it services five 154 kV circuits and two 66 kV circuits. Damage to a 154 kV pantograph disconnect switch was reported. A pipe near the Buchholz relay failed. Rigid bus supports on 15 busses on top of a 66 kV rack failed.

Figure 6.124 shows part of a failed substation wall, suggesting that the ground motion were relatively low. Figure 6.125 shows cracking in the ground indicating lateral spreading; however, no cracking was observed near the equipment in the switchyard. Figure 6.126 shows a transformer on an elevated pedestal, which contains a catchment for oil in case there is a leak. Note that the conservator is high above the tank, but its support structure is braced. Figure 6.127 shows what appears to be a vent pipe adjacent to the Buchholz relay, but there is no connection. The vent pipe was close to the pipe connected to the Buchholz relay, and impact loading may have contributed to the failure. A short section of pipe was replaced because of a weld failure. It is not known if the conservator support structure had slotted connections that slipped, as observed at the Charrua substation.

Figure 6.123. An interesting detail, called a pig tail, provides slack between equipment

Figure 6.124. Several sections of substation wall collapsed

Figure 6.125. Cracks in ground probably from lateral spreading

Figure 6.126. Transformer on pedestal that contains a catchment for oil if there is a leak

Figure 6.127. Damaged pipe connection may be due to interaction with riser

Figure 6.128. Deformed pigtails indicates moment of equipment

Figure 6.128 shows the use of pigtails to provide slack in conductor between equipment. Note that the pigtails in the right foreground have been kinked or opened, indicating that there had been movement, while those in the background are not deformed.

There were several pantograph disconnect switches in the switchyard. Some damage was experienced, but it could not be identified. Λ rigid bus on largc racks was damaged; a similar failure mode will be shown in the SE Coronel substation.

6.5 Distribution System

Summary of Distribution Substation Damage

Table 6.4 summarizes damage to distribution substations.

Detailed Review of Distribution System Substation Damage

SE Coronel Substation

The SE Coronel Substation is located south of the Bocamina generating station near the coast and is serviced by one 154 kV circuits and five 66 kV circuits. Three 154 kV candlestick circuit breakers and two surge arresters failed. An unanchored power transformer may have moved with no consequence. A transformer anchorage detail at distribution substations has allowed slipping and possible lifting. The T-bar ceiling supports in the control room were damaged and a plastic electrical conduit broke, but the electrical connection appeared to be intact.

Figure 6.129 shows a section of fallen T-bar ceiling support. All of the ceiling tiles had been removed. The connection of plastic conduit failed at a junction box, but the electrical connection appeared to be intact (Fig.6.130). A suspended cable tray supported the conduit, and the motion of the tray failed the conduit.

Figure 6.131 suggests that an unanchored transformer moved, but with no consequence. Figure 6.132 shows an anchorage detail seen at distribution substation. This design is relatively flexible to uplift and was deformed. A similar anchorage was torn at the SE Hospital substation and is discussed later. It is unclear if there was any vertical deformation in the anchorage or if there was a gap when it was installed. The transformer slid about 25.4 mm (1 in.).

Figure 6.133 shows three failed 154 kV candlestick circuit breakers and a 154 kV surge arrester. Another surge arrester was found elsewhere at the site. Figure 6.134 shows a common 66 kV rigid bus attached to the top of a frame; in this case the bus bars are missing. This type of damage was reported at several substations. In some cases the bus bars were in place, but the supporting insulators were damaged (Fig. 6.135 and 6.136).

Figure 6.129. A small portion of the T-bar ceiling support fell

Table 6.4. Summary of Distribution Substation Damage

Substation	Operating Voltage			Miscellaneous, Comments
	220 kV	156 kV	66 kV	
SE Coronel 154 kV, 66 kV		One to three candlestick CB failed SA failed	7/15 busses on a rack had failed supports	An unanchored TR may have moved T-bar ceiling failed, tiles failed Plastic conduit
SE Hospital 66 kV 0.35 g			Double -break DS control linkageTR internal damage TR anchorage, low voltage bushing leakTR low voltage SA terminal Bus support structures moved Distribution TR leads	Dead-end structure foundation showed that there was structural movement
Paine 154 kV, 66 kV		3 air blast, LT CB		Storage cabinet blocking door
Quilicura 110 kV, 12 kV			Six 100 kV candlestick CB	
Substation view from highway 220 kV	4 candlestick CB			
Inforsa		2 TR bushings		
Itahue		Air blast CB		
Mapal		1 TR bushing		
Fopaco		2 TR bushing		
CAP (Steel Plant)		9 TR bushings		
CMPC		6 TR bushings		
Talcahuano			DS failed	
Uranilahue Huachipato			TR failed	
Parral			Rigid bus top rack	

Figure 6.130. Damaged electrical conduit

Figure 6.131. Track suggests that the unanchored transformer moved

Figure 6.132. An anchorage detail found that allowed sliding and possible lifting

Figure 6.133. Three failed 154 kV candlestick circuit breakers and a surge arrester

SE Hospital Substation

The SE Hospital substation is located south of Santiago and is serviced by two 66 kV circuits. The water table at the site is less than 1 m below the surface. The most significant damage was to the low-voltage transformer bushing caused by motion of bus supports, its anchorage, and internal damage. A 66 kV double break disconnect switch was damaged. There was an unanchored communications rack, but no damage resulted.

An unanchored storage rack tipped over and jammed the door closed. Tears in transformer anchorage hardware indicate that there was significant ground motion at this site.

Figure 6.134. Rigid bus (missing) on top of frame supported bus insulators

Fig. 135. Damaged insulator supporting bus

Fig. 136. Damaged insulator supporting bus

The most significant damage at the substation was to the 66 kV transformer. Figure 6.137 shows the low-voltage bushings on the damaged transformer. In the background overhead, a bus with drops to the low-voltage bushings on the replacement transformer can be seen; the view of which is partially blocked by the conservator of the damaged transformer. The overhead bus is connected to the support structure on the left, identified by the upper left arrow. The right support of the bus is just out of view on the right. The drop from the overhead bus passed through an eyebolt connected to the surge arrester terminal. The lower arrow on the left points to the surge arrester associated with the B-Phase. The A-Phase (out of view to the left) terminal hardware was damaged, which is associated with the damaged bushing. The two structures supporting the bus tipped away from each other due to moment of their foundations, thus causing the drop to be pulled up as the overhead bus became taught. The eyebolt failed, the tension on the bushing bent the terminal, as shown, and the bushing developed a leak. Figure 6.140 shows the replacement transformer, but the low-voltage bushings are in a protective crate and the surge arresters have not yet been installed. The internal circuitry of the transformer was damaged. This is one of four distribution transformers that had internal damage, three of which are located at unidentified substations.

The anchorage details on the old and replacement transformers are the same, and two beams can be seen at the base of the transformer (Fig. 6.138). Figure 6.139 shows hardware in which the box bolts to the transformer base beam with a connecting plate, which is used to secure the transformer to the foundation. The plate is bent, and there is a 50.8-mm (2-in.) tear in the weld that connects the plate to the box. The length of the base plate between the bolts and the connection to the box assembly provides a flexible connection. A second anchor assembly also had a cracked weld, but the damage was not as severe.

Figure 6.137. Damaged low-voltage bushing with bus connections in background

Figure 6.138. Replacement transformer being installed

Figure 6.140 shows a double break disconnect switch mounted on top of a dead-end support structure. Two of the phases are open, and one is closed, indicating that there was probably a failure in the linkage controlling the switch phases. Figure 6.141 shows new connections to several single-phase voltage regulators that used pigtails to provide more slack. One or more of the connections had a problem, but there was no equipment damage.

Figure 6.142 shows an unanchored communication equipment rack and largely empty storage shelves on the left. At the time of the earthquake, the door was closed and the storage shelves tipped over preventing the door from opening. The cracks in the door were caused by an attempt to force it open.

Figure 6.143 shows a perfectly executed installation of the station batteries. The rack is well anchored, the restraints holding the cell to the rack fit tightly, and there are foam spacers between cells to prevent the cell movement from stressing terminals. Additionally, there is flexibility in the connections between the terminals and the wall connection. The third battery from the end on the bottom rack did slip slightly, but it was not damaged.

Figure 6.139. Torn weld on transformer anchor connection hardware

Figure 6.140. Damaged linkage on double break disconnect switch

Figure 6.141. Pigtails replaced short lead on the single-phase voltage regulators to increase slack

Figure 6.142. Unanchored communications rack and storage shelves blocking the door

Figure 6.143. A well-executed station battery installation

Paine Substation

The Paine substation is located just north of the SE Hospital Substation and is serviced by one 154 kV circuit and three 66 kV circuits. A 154 kV air-blast circuit breaker failed. Damage was limited to older equipment that was known to be vulnerable, suggesting that the ground motions were moderate.

Figure 6.144 shows damaged parts to a 154 kV air blast, live-tank circuit breaker outside the substation with the circuit breaker tanks in the background. The damaged unit was being replaced with a new candlestick circuit breaker.

Figure 6.145 shows a 66 kV air-blast circuit breaker. The unit is relatively small, but it is mounted on wheels that are restrained. Slack to the adjacent units is limited.

Figure 6.146 shows a 154 kV to 66 kV transformer, which is secured with plates that act as hold down dogs. Figure 6.147 shows one of the four hold-downs used to secure the transformer. A pin secures the transformer base plate to the hold down plate. The transformer did not move.

Figure 6.148 shows marks left on the wall from the earthquake-induced rocking of a fire extinguisher.

Figure 6.144. Broken parts to a 154 kV air-blast, live-tank circuit breaker

Figure 6.145. A relatively small air-blast circuit breaker installed on wheels

Figure 6.146. A 154 kV transformer secured with hold down plates

Figure 6.147. One of four plates anchoring the transformer

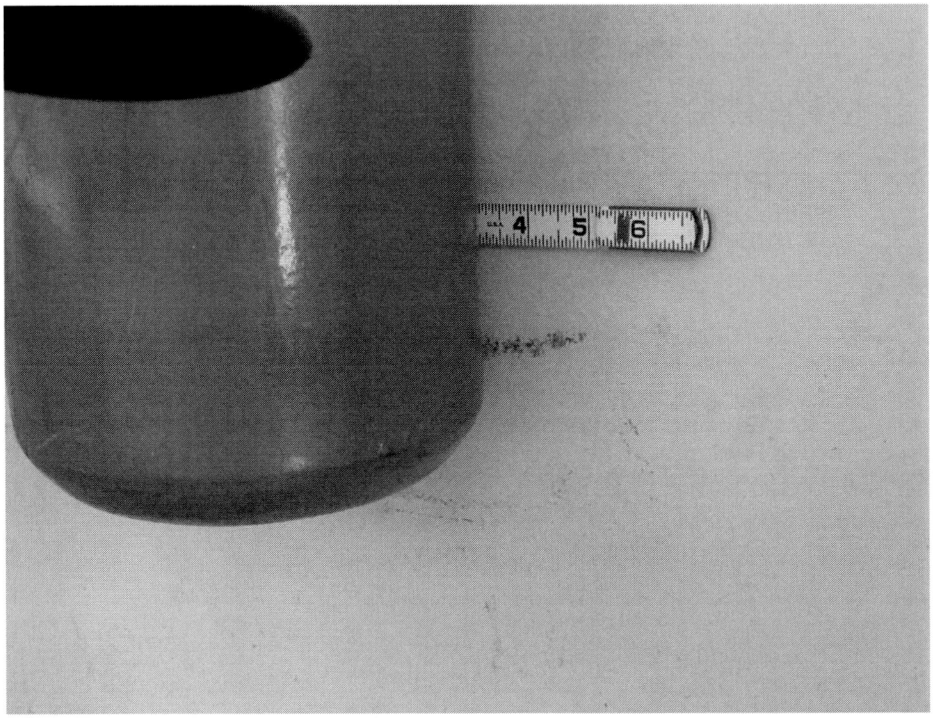

Figure 6.148. Marks left on wall due to earthquake induced rocking of fire extinguisher

Quilicura Substation and Unnamed Substation

The Quilicura substation, which the team had to view from the gate, is located north of Santiago. It is a 110 kV to 12 kV substation and appeared to supply an industrial facility. Most of the site had not been cleaned up, and it is assumed that the industrial facility had problems and was not in service. There were approximately six 110 kV candlestick circuit breakers on the ground. Some post insulators, which may have been bus supports or parts from disconnect switches, were also on the ground.

Figure 6.149 shows three failed 110 kV candlestick circuit breakers in the foreground and a fourth in the background. There were at least two other candlestick circuit breaker failures in other parts of the substation. Several post insulators were also observed on the ground, but it is not known if these were from bus supports for disconnect switches.

While driving north from Santiago to the Polpaico substation a photograph of another substation was taken from the highway (Fig. 6.150). After the trip the photograph was examined closely, and there appears to be four damaged 220 kV candlestick circuit breakers on the ground.

Figure 6.149. Damaged 110 kV candlestick circuit breakers in a distribution substation

Figure 6.150. Damage in substation observed from highway

6.6 Transmission and Distribution Lines

Transmission Lines

Damage to transmission lines was to three towers lost at a river crossing near the mouth of the Biobío River.

Figure 6.151 shows the location of the power line river crossing near the mouth of the Biobío River. Each of the three towers was supported on extended piles to keep the tower above the flood stage of the river. The height of the piles above the river on the second tower appears to be about 5.8 m (19 ft.) above the riverbed, which was exposed when the pictures were taken shortly after the earthquake. Other towers that appear to be founded out of the river channel were not damaged. Because the crossing is so near the mouth of the river with the Pacific Ocean to the west, the base of the tower could have been subjected to brackish water by tidal activity or affected by the tsunami. Figure 6.152 shows the southernmost tower of the three that failed, which had fallen to the east. Parts of two of the steel foundation piles that were on the tower's west side wound up on top of it, and the painted part of the pile was above grade. Note that it appears that this tower is only supporting a single circuit (three conductors) and a lightning shield wire on top, while the other damage towers carry two circuits. It has been reported that most of the towers carried two circuits (154 kV and 220 kV), but only the 154 kV circuit was in use. However, the towers to the south also carried two circuits. This figure shows that one of the piles, which is now on top of the fallen tower, broke a few feet below the pile cap, and the rest of the pile appears to be its original position in the ground. The second pile fractured a few feet below the ground surface, and it is assumed that most of the end of the pile was left in the ground. This figure appears to show that the tower steel is largely undamaged.

Figure 6.153 shows a picture of the fractured end of the pile that failed below the ground surface, which appears to be a corroded steel pipe filled with concrete. The character of the failures suggests that they occurred at welds in the pipe.

Figure 6.154 shows a closer view of the base of the tower. The upper side of the fallen tower appears to be undamaged and clean of water-borne debris. This suggests that if the tower was subjected to the action of the tsunami, the height of the tsunami was below the height of the top of the support piles. In addition, the plants and small trees at the base of the tower appear to be undisturbed; thus, presumably the base of the tower was not subjected to the action of the tsunami. What were the east legs of the tower cannot be seen; it is assumed that there was liquefaction, and they sunk into the soil. It appears that only the east legs sank, which could be due to eastern pull on the tower increasing the load on the eastern piles and reducing the gravity load on the western piles. The liquefaction potential can vary by location, which could explain why the west piles did not sink. Note that the lower legs (east) of the tower are deformed, which could have occurred when the east side of the tower dropped as the piles sank in the liquefied soil. The buckling at the base of the east tower legs could have caused the tower to lean to the east, which applied an added load to the east piles and reduced the gravity load on the west piles.

Figure 6.151. Map showing river crossing (Courtesy of Transelec)

Figure 6.152. Fallen tower with failed foundation piles (Courtesy of Transelec)

Figure 6.155 shows the two damaged towers to the north. Note that the tower in the foreground supports two circuits. The conductors have different diameters, but the insulators are the same length. Transelec indicated that the crossing supported two circuits: one 154 kV and one 220 kV. They used the same size string insulator for both voltages, as larger insulators were to be used for a latter upgrade to two 220 kV circuits. Note that an undamaged tower to the north is supporting two circuits.

Figure 6.156 shows two replacement towers with one circuit. It appears that the supports for the second circuit on the towers to the north have been removed. Two conductors can be seen in the foreground, which may be from the unstrung circuit. A shield wire can also be seen; however, it appears to be missing on the replacement towers.

It is often difficult to assess the sequence of events with a cascading failure such as this. The duration of the sinking of the piles on the south tower due to liquefaction is long compared to the dynamic variation in the overturning moment on the tower due to earthquake excitations. If the tower to the north had failed first, it could have pulled on the lines causing a long-term moment on the foundation, thus reducing the gravity load on the western piles and adding load to the eastern piles on the south tower. The differential loading could have caused the eastern piles to sink. Alternatively, if there had been different liquefaction potential at the piles of the south tower, the east piles could have sunk, causing the tower to tip, which could have pulled down the towers to the north.

Figure 6.153. Broken tower pile in weld area (Courtesy of Transelec)

Figure 6.154. Eastern tower legs buckled (Courtesy of Transelec)

Figure 6.155. Two towers to the north were damaged (Courtesy of Transelec)

Figure 6.156. New towers restoring one circuit

On the south tower, as the eastern piles sank due to liquefaction, large bending moments were applied to the western piles causing them to crack. Additionally, as the east side of the tower sank, large moments were applied to the tower's east legs, causing them to buckle.

Distribution Lines

Several hundred concrete distribution poles were damaged, many in areas impacted by the tsunami. There are indications that the loss of poles in the tsunami may be due to debris carried with the water rather than from the force of the water directly. Damage to wooden poles used in the Unites States is rare, and it is usually due to landslide or soil failure at the foundation. The cable loading on distribution poles in Chile is much higher than in the United States.

A very small percentage of the total inventory of concrete distribution poles were damaged—1,591 out of 759,186. About half were in the tsunami inundation zone, where many of the facilities were severely damaged. Figure 6.157 shows an area adjacent to the shore in which a distribution pole is slightly tilted, but still standing. Note that a structure in the background has been stripped of its cladding. Many poles in the area were down. It is speculated that the downed poles were struck by debris carried in or out in one of the three major tsunami flows. One post was left standing. A wooden pole indicated by an arrow is shown in the left of the figure. It is not clear what function this pole served, as all utilities poles appeared to be concrete. Figure 6.158 shows that this pole survived, even though the incoming waves would have a direct shot at it. Figure 6.159 shows two poles that supported a distribution transformer. Again, this site had a direct view of the coast, a couple of hundred meters from the shore. The transformer could not be found, and there is a possibility that it was looted for its copper. However, other units nearby were undisturbed. Thus, the poles could withstand the loads of the water tearing the transformer from its supports without being uprooted.

In urban areas, poles were heavily loaded, as is illustrated in Figure 6.160. A few of these poles were downed by cable pulls from façade wall collapses.

Misalignment of the insulators on a pole near Highway 5, which links Los Angeles with Talca, suggests that there was a problem with the pole or its foundation (Fig.6.161). Figure 6.162 shows the base of the distribution pole is buckled. It is not known if this damage is earthquake related. Additional information on the structure and transmission parameter would be needed to determine the cause of failure. One of Transelec substation personnel mentioned that this was not earthquake damage but a result of a broken conductor caused by copper wire theft.

Figure 6.157. Distribution pole survived the tsunami with a slight tilt

Figure 6.158. Wooden pole survives the tsunami

Figure 6.159. Tsunami stripped poles of their transformer, but they survived

Figure 6.160. Poles in urban areas can be heavily loaded

Figure 6.161. Misaligned poles suggests pole or foundation problem

Figure 6.162. Buckled pole base

6.7 Conclusions and Observations

Earthquake Magnitude and Ground Motion

- This earthquake had a large magnitude and long duration ground motion.
- The limited evaluation of strong motion data indicates that spectral acceleration amplification frequently greatly exceeded values predicted by the required response spectra used for equipment qualification.
- Peak accelerations at substation sites were not estimated because of the large spectral acceleration as compared to the peak acceleration made it difficult to apply experience from past earthquakes to this event.
- The duration of shaking was much longer than has been experienced to date in the United States, but long duration excitations can be expected in future large and great earthquakes.
- The intensity of shaking over much of the region in this subduction earthquake was much less than experienced in crustal earthquakes in the United States.
- The ground motions at one site in this earthquake suggests the need for IEEE 693 to require sites that use base isolated equipment be evaluated for their susceptibility to large, low frequency ground motions, as qualification to the RRS does not assure acceptable performance.

General Observations about Power Systems

- The use of cell phones for the dispatch of crews for substation assessment and restoration was problematic and is inherent in the technology as it is power dependent. Radio-based systems are not extensively used, as is the case in the western United States.
- Power dispatch needs a more robust SCADA system, which was recognized before the earthquake and is being addressed.
- Damage to transformer bushings appeared to be disproportionately high and duration of excitations may have had an influence.
- Long duration of shaking did not demonstrate that it affects substation equipment with possible exception of transformer bushings.
- The shape of the country, the evolution of substations and their equipment are impediments to reliability. There is a need to implement N-1 reliability to the entire system and not limit it to transmission lines. The backbone 500 kV system in the western United States may exhibit similar characteristics, such as lack of transmission line route dispersion and a system that is largely linear in character.
- The three types of equipment that are used sparingly in the United States, pantograph disconnect switches, live-tank candlestick circuit breakers, and low-voltage, air-blast circuit breakers appeared to be the most vulnerable. A relatively small percentage of the inventory of this equipment was damaged. The low-voltage circuit breakers were old and being phased out.

- While seismic qualification requirements in Chile are less stringent than those in the North America, they generally appear to be adequate to meet Chile's needs, with possible exception of transformer bushings.
- Damage to limited thermoelectric generating sources in the south has required the use of less than optimal economic dispatch, but the construction of new plants would address this issue.
- The investigation team had unprecedented access to facilities and there was open disclosure of damage data.

Transmission System Response

- Good seismic practices were implemented, such as anchorage of equipment, adequate slack in conductor between equipment in switchyards, and installations of station batteries.
- Power restoration times were fast.
- In the majority of cases, equipment damage was concentrated in circuit breakers without damping devices and in pantographic disconnect switches with the upper contact supported by post insulators.
- Extensive use of damping devices on many types of high voltage equipment, particularly dead-tank circuit breakers, enhanced response. The use of damping devices in the western United States has been primarily done on an experimental basis.
- The only transmission tower damage was at a river crossing. Soil liquefaction and a defective pile weld contributed to the failure of one tower. Other towers were damaged as a result of the failure of the first power. For critical circuits, pre-design site evaluations are recommended because of the time to restore this type of tower.

Distribution System Response

- Substation busses were generally energized within 1 to 3 days
- Restoration of service to at least 99 percent of the customers was generally done within 10 days, but extended to two weeks for some.
- Damage to concrete distribution poles was more severe than to the wooden poles used in the United States, but as a percentage of their inventory, damage was very small. Poles in Chile are more heavily loaded with communication lines in urban areas than in the United States.
- As is the case after earthquakes in the United States, details of the issues with the distribution system feeders was not investigated.
- The tsunami was very damaging to power system infrastructure, but it may be due to debris carried in the water rather than the direct action of the water.
- There was much more damage to equipment in substations operating at 138 kV and below than is observed in the United States. This is probably due to the greater use of candlestick and air blast circuit breakers in Chile.

6.8 Acknowledgements

The author gratefully acknowledges the preparation of special plots by Professors Boroschek and Contreras for selected earthquake records with over plots of IEEE 693 Required Response Spectra.

6.9 References

Chilean legislation in 2009, “Norma Tecnica de Seguridad Calidad y Servicio” established acceptable seismic qualification provisions.

Boroschek, R., Soto, P., Leon, R. "Maule Region Earthquake, Feb 27, 2010, Mw = 8.8" Renadic Report 10/08 Rev. 1, October 2010.

7
TELECOMMUNICATION SYSTEMS

Collecting telecommunication performance information after an earthquake proved once again to be a challenging task. As always, persistence helped to collect relevant information for future improvement of both design and installation practices for telecommunication equipment and for network layout to develop a more resilient system. The service providers in this industry are extremely competitive. Cost-benefit and risk management are the two key factors in sustaining the corporation and driving advances in technology.

Like many Latin America countries, modern telecommunication in Chile began with the privatization of the state-owned companies in the late 1980s and early 1990s. In 1989 the two main communications companies in Chile were CTC, which provides plain old telephony service (POTS), and ENTEL, which provides long distances services. Both were acquired by Telefonica of Spain, although it sold its participation in ENTEL in the mid-1990s. Thanks to the deregulation process that started about that time, the service providers sped up the network growth to satisfy the increase in demand. The Chilean telecommunication system was also influenced by the United States, particularly during the last two decades. AT&T and Bell South were among the U.S. telecommunication

Figure 7.1. A manhole cover spotted in Santiago with the AT&T name on it

companies that entered this market and were responsible for setting up much of the network architecture, so hardware remnants show their presence (Fig. 7.1).

The major companies competing in this market are Telefonica (under the new market name of Movistar), ENTEL, Nextel, and Telmex (under its South America brand name of Claro[1]). They all provide cellular[2], landline, and Internet[3] services but under a different brand name. VTR the largest cable company in Chile, is one of the three companies that provide TV service, and VTR also provides Internet and landline services. The other television providers are Telmex and Telefónica. Telmex is a relatively new comer in this market, although the part of its network purchased from Chilesat had been in operation since the late 1980s. We are grateful to both Telefónica and ENTEL (in Concepción) for accepting our interview and providing relevant information and photographs of earthquake damage for this report.

The overall performance of the telecommunication service in this earthquake was moderate to poor, depending on the area. The cellular network performance was poor comparing to the landline network. This may be a result of many utilities using the cellular phone system as their primary maintenance and service crew dispatch and coordination communication tool. We were surprised that the airport operations staff uses cellular phones as their daily communication tool as well.

It was not surprising that unanchored equipment did not perform well. A failure mode now identifies a new vulnerable element in the cellular system—antenna failure. More antennas were damaged than had been observed in other major earthquakes. Two items reinforced our belief that change is needed to improve the post-earthquake performance of the cellular system—backup (or reserve) power and base transceiver station[4] (BTS) using commercial or non-telecommunication equipment buildings to house equipment and antennas.

The long and narrow geographic shape of the country prohibits optimal dispersion of the network to provide true redundancy. A north-south link along the foothill close to the mountain range where seismicity is lower means a more costly connection to the populated areas along the coast. However, links can be developed along the low-seismicity areas as the demand for service increases. Implementing a risk reduction plan along with the overall market share growth plan will result in better post disaster performance.

---

[1] Claro was rebranded from CTI Móvil since March 2008. It is owned by América Móvil, a Mexican telecom group.

[2] Cellular, wireless, and mobile phone are used interchangeably. The term cellular is used through out this chapter.

[3] Internet is sometime referred to as Information Technology (IT). Internet is used thorough out this chapter.

[4] BTS are physically located in so-called cell sites. Sometimes BTS and cell sites are used in an indistinguishable way. For consistency, BTS is used in this chapter.

7.1 Description of System

The telecommunications network in Chile is similar to that in the United States, both in terms of hardware and architecture. The hierarchical central office (CO) structure yielded to a flat architecture with the introduction of digital switching and Common Channel Signaling System 7 (CCSS7) call processing protocol. Although in most cases the tandem offices retain the same name, their functions are more than just routing long distance calls.

In addition to optical fiber cables for connecting COs and remotes switches (RSs) both copper cables and microwave are also used. Microwave is used mainly for reaching locations where either optical fiber or copper cable routing is extremely difficult and costly. Remotes switches are very small COs used to serve a rural area with less than 5000 customers.

Since the capacity of a fiber cable is many times that of a copper cable of similar physical size, the failure of a fiber cable will affect more customers. For example, the conventional T3[5] copper cable has a capacity of 44.736 Mbps, while an OC48[6] fiber has a capacity of 2,488.32 Mbps. That is the optical fiber cable has a capacity of more than 50 times that of copper cable.

A significant difference from U.S. communication networks operation, particularly in restoring service after disasters, is that in the United States there is some degree of shared infrastructure among network operators, particularly in cell sites where antennas from different operators are placed on a same tower, while, in Chile shared infrastructure among network operators is minimal. For example, cell sites are not shared space. However, the electric distribution infrastructure is shared with the wireline communications network infrastructure in both countries.

From an emergency electric power perspective, most large COs have power generators set up along with battery reserves to keep equipment operational during power outages (Fig. 7.2). The battery reserve in most COs is around 8 to 10 hours. Most RSs do not have back up power generators, and their reserve power is only for a few hours. The same is true for BTSs. In some instances, the battery stands are not anchored, and the batteries are not secured to the rack properly.

Cellular phones are becoming the most popular telecommunication tool, not only for the public but also for utilities using it as their service dispatch tool. This earthquake proved the later to be a mistake. Similar to North America, most BTSs are located on commercial building rooftops. Collateral damage caused by building failure was quite common. Figure 7.3 shows an example of colocating and concentration, but this ENTEL tower was built for telecommunications, and this is not a commercial building. Figure 7.4 shows a BTS with its antennas installed on the roof of a commercial building.

[5] T3 = T-carrier 3 (digital transmission line, 44.736 Mbps, 672 voice channels), also known as 3 T1 trunks

[6] OC48 = Optical Carrier 48

Figure 7.2. The emergency power generator in a Concepcion central office

Figure 7.3. A concentration of antennas and microwave dishes collocated in a tower

Figure 7.4. BTS installed in a commercial building

Both the landline and cellular network equipment are totally digital. Cellular networks are keeping pace with the technology advances to 3G providing data, Internet, and video streaming services. The telecommunication equipment was sourced from manufacturers from all over the world.

The network configuration is not much different from those in the United States. It consists of optical fiber loops with digital loop carriers (DLCs) and/or digital subscriber line access multiplexers (DSLAMs) remote terminals (RTs) on utility poles, elevated platforms, or on the sidewalk. These DLCs and DSLAMs have usually a very limited battery power backup and are powered from power lines sharing poles with RT distribution output cables. The network provides connectivity to all landline and cellular services such as voice, data, Internet, and such. The telecommunications network is now so complex that a cloud is often used to represent the physical connections and the services that reside in different equipment offices (Fig. 7.5). In fact, cloud technology started from telecommunications when virtual private networks[7] (VPN) bccamc available in the early to mid-1980s.

The television cable network is now part of this telecommunications network, as the cable companies are also entering into this lucrative market providing the traditional television services as well as Internet service including Voice over Internet Protocol (VoIP).

[7] Trade names such as Centrex are also used.

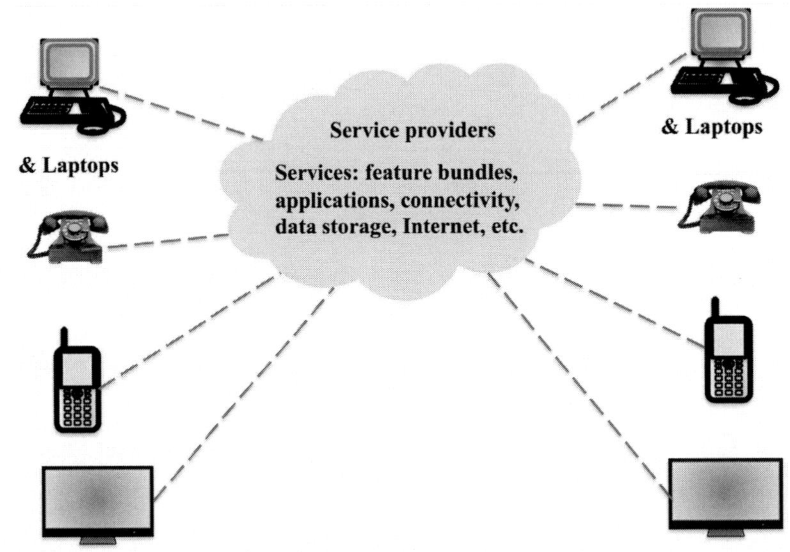

Figure 7.5 The convergence of telecommunications networks can be represented by a cloud.

The advances in personal digital assistant (PDA) technology and design, particularly with the reduced size of high-capacity batteries, have resulted in a high demand of network capacity. The increase of Wi-Fi nodes allowing faster and easier entering points is fuelling the speed of reaching the network limit.

Plain old telephone service (POTS) has really changed in the past two decades. The performance of this network (or convergence of telecommunications networks) must be evaluated and standards set to ensure the reliability of this critical lifeline service.

7.2 Overview of System Performance

The performance of the telecommunication system in the earthquake-impacted area was moderate to poor. The damage observed was distributed; there were no one or two heavily damaged sites. The reported and observed damage are:

- equipment failure (Fig. 7.6),
- batteries fallen off racks (Fig. 7.7),
- severed cables colocating with bridges (Fig. 7.8),
- backup power generator failure, and
- tower and antenna failure (Fig. 7.9).

In addition, electric power outages had the greatest impact on the network performance. Both landline and cellular services were mostly restored 7 days after the earthquake. Restoration in a few areas within the most affected regions of Bio Bio and Maule took up to 2 weeks. There were many factors affecting the recovery duration.

Figure 7.6. Equipment on raised floor damaged, Concepción

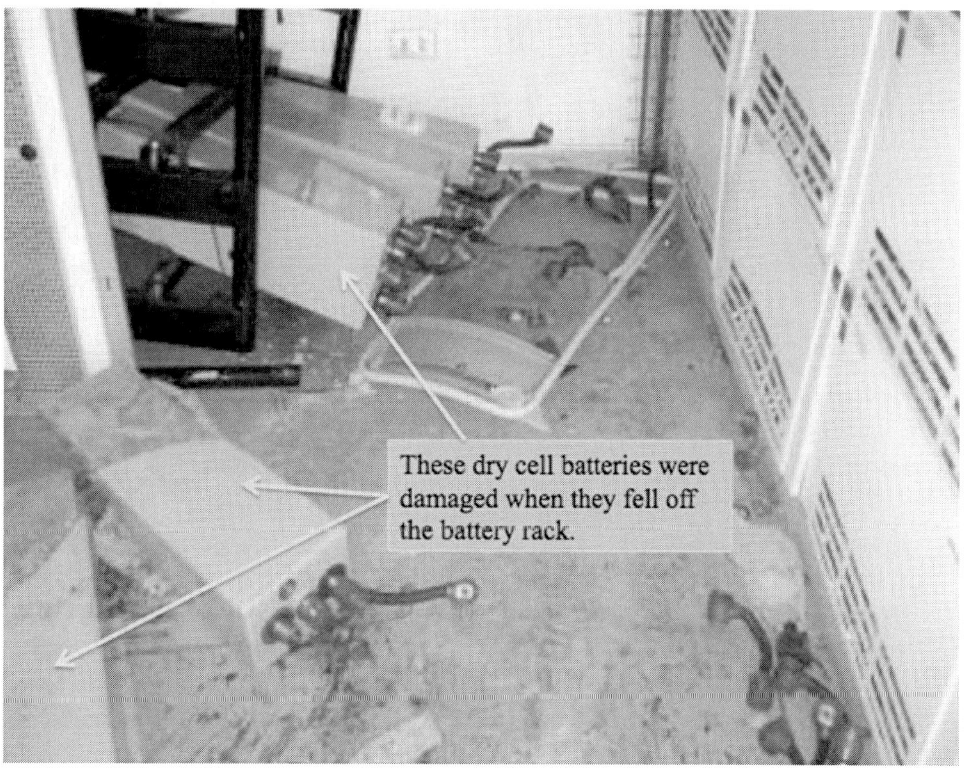

Figure 7.7. Damaged batteries that fell off the battery rack (Photo reproduced with permission from its source)

Figure 7.8. Colocated cable damage at Puente Juan Pablo Segundo Bridge

Figure 7.9. Cellular antenna mounting pole anchor failure
(Photo reproduced with permission from its source)

This report does not detail the system damage within the tsunami-impacted areas because these areas were demolished by the tsunami. In communities that survived the tsunami, wireless service was restored within 5 days by one of the service providers free of charge.

The transmission and distribution networks sustained minor damage in this earthquake. The damage to the distribution system was mainly from building failures that pulled the drop wire from the utility poles (Fig. 7.10). One service provider had to realign 10 microwave dishes (Fig. 7.11) and also had a long haul with a few optical fiber cables severed. However, due to shared access in the post deregulation environment, long distance calls were not severely disrupted because traffic in damaged cables could be routed through undamaged cables nearby.

The most significant damage occurred in the cellular network. More than 40 percent of BTSs operated by one of the service providers in the earthquake-impacted areas had antenna failures—300 in Region VII, and 400 in Region VIII. This totaled about 1700 BTSs. Verbal reports from users indicated that all cellular service providers in the area had the same interval of service disruption. The amount of damage to these service providers in these regions was probably about the same.

The service providers we met had backup power generators installed in critical facilities, such as large switching offices and transmission offices. However, most offices with fewer than 5000 subscribers lack generators. In total, approximately 150,000 landline subscribers lost service due to lack of power in these small central offices. In larger central offices, backup power generators were either installed inside the office or in a separate building outside the office. The fuel supply onsite was good for 2 to 3 days. Except for two incidences, all the backup power generators started up without any problem. One incidence was caused by a transfer switch failure; the other was due to a fire within the generator room. One of the failed backup power generators was used to power the air conditioning equipment; the switching equipment shut off due to overheating.

The fuel supply was one of the main power supply issues. Supply disruptions resulted from blocked road access, refinery shut downs, power outages, and pilferage. As there were insufficient small power generator sets, only about 10 percent of the BTS sites had generator sets installed (Fig. 7.12). Some of these power generators were provided by Argentina through a mutual emergency support agreement. Fuel and generator sets were stollen 2 to 3 days after the installation.

Water departments, electric power companies, airport operations, and other government organizations that utilize cellular services as their operations communication tool experienced difficulties in the post-earthquake emergency response and recovery processes. All of the organizations visited by the ASCE/TCLEE team members indicated that using a separate radio system as a backup or a primary communication tool instead of relying on cellular phones would provide improvement right after a disaster.

Figure 7.10. Distribution system: drop wire severed due to building damage

Figure 7.11. Inertial load caused the microwave dish to rotate out of alignment

Figure 7.12. Small power generator units installed outside a BTS in Constitución

Landline telephony service recovered faster than wireless coverage. This occurred because most network outages were caused by a lack of power in mostly RSs for landline service and in BTSs in wireless networks. As the number of affected BTSs was significantly higher than the number of affected RSs, power restoration was simpler for the landlines than wireless service. Still, one report indicated that after cellular service was restored, the landlines went out for a day in some areas within Regions VII and VIII. The service providers are still working to discover the cause of the problem and prevent it from happening again. As with many previous disasters, network congestion significantly affected all network users. At the time of the earthquake, Chile lacked a call prioritization service, so all users—even emergency responders and government officials— were equally affected after the disaster.

7.3. Cellular Networks

Cellular service providers in regions VII and VIII sustained heavy damage in addition to the electric power outage. Particularly in the remote areas, access to provide power or to restore damaged equipment became extremely difficult due to damage to roads and bridges. The dry season reduced the number of landslides, which could have made access worse.

We observed a wide range of cellular tower designs from Santiago to Concepcion. In cities such as Santiago, many BTSs are located on rooftops of tall buildings. The value of land also limits BTS locations within the city to provide proper signal coverage. Figures 7.13 to 7.16 show the different types of rooftop installations observed.

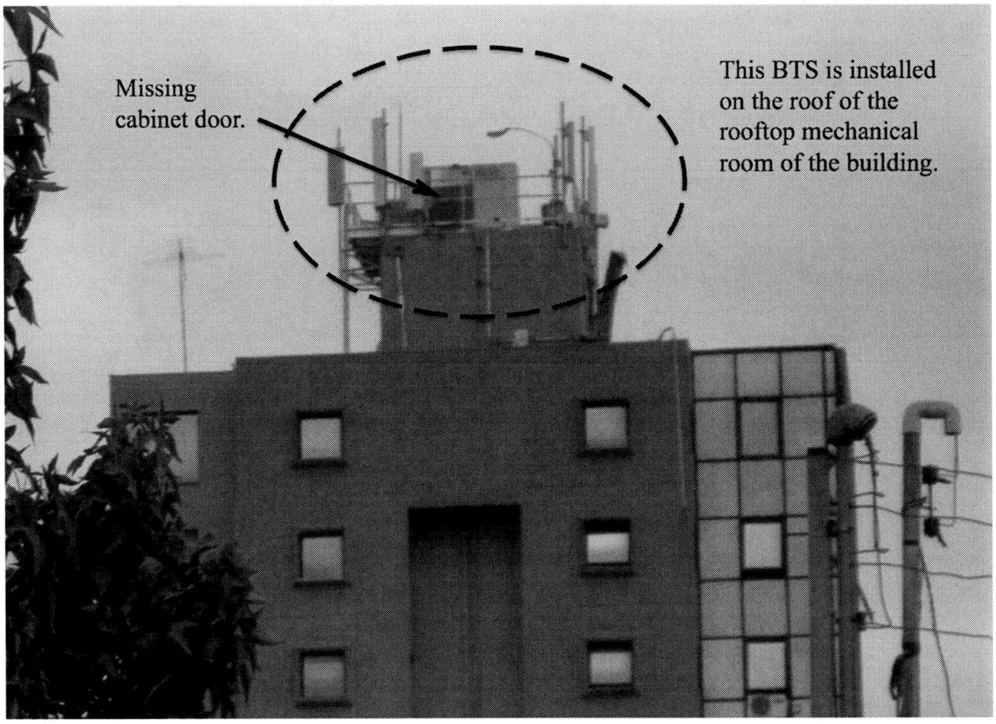

Figure 7.13. BTS installed on the rooftop of a commercial building in Curico, south of Santiago. (S34.983069°, W71.238876°). Notice the fallen cabinet door indicating potential damage to the BTS due to shaking

Figure 7.14. BTS installed on the A-frame roof of the silos in this manufacturing facility south of Santiago. The second container at the right at the Silo collapsed from the earthquake.

Figure 7.15. BTS installed on a rooftop

In rural areas, many BTSs are installed on a property enclosed by chain link fence or concrete wall fence with stand-alone towers of three main types. They are unipole (Fig. 7.17), cable guided (Fig. 7.18), and traditional lattice design (Fig. 7.19). These towers were designed and installed based on wind load. Although a couple of towers collapsed during the earthquake, in general they all performed well. Further investigation will identify why these towers collapsed.

In populated areas, service providers usually locate BTSs close to each other. Figure 7.20 shows three service providers within a few meters of one another. Service providers locate the BTSs close together and along the highway between cities, particularly major highways and those with heavy traffic flow. A hilltop overlooking a major populated center is another location used by cellular service providers (Fig. 7.21).

The ASCE/TCLEE members relied heavily on service providers to provide the team perishable data and photographs of damage. We are grateful to both Telefonica and ENTEL for their support of our effort to reduce future earthquake loss and improve performance of telecommunication systems.

Figure 7.16. BTS colocated with a wireline central office. Due to space constraints, the tower was building on top of the roof to gain coverage among high-rise buildings in the area.

The failure modes of the cellular networks are:

1. antennas fallen from towers,
2. BTS failure due to colocating,
3. antennas tower failure,
4. batteries fallen from racks,
5. equipment failure, and
6. power outage.

The following sections summarize the performance of the sites visited and the sites reported by the service providers.

Figure 7.17. Unipole style cellular antenna towers

Figure 7.18. Cable guided tower

Figure 7.19. Traditional stand-alone tower design

BTS Cellular site (S36.890184^0, W73.137965^0)

This site is located southwest of Concepcion (Fig. 7.22). The collapsed exterior concrete wall showed that this site experienced strong ground motion. A steel chain link fence around the BTS provided additional security. This site was decommissioned, most likely due to the extensive damage to the equipment inside the building. Figure 7.23 shows that the waveguide cables were cut off, and most of the radio equipment was removed from this station.

The power equipment (Fig. 7.24) was installed outside the building on its own concrete pad and anchored. The batteries were installed on a rack inside the building; however, they were not secured on the shelf to prevent them from falling off during strong shaking (Fig. 7.25). The battery rack, which was not anchored, rotated about 12 cm (4.5 inches) from its original position (Fig. 7.26).

Figure 7.20. Three cellular service providers locate their BTSs within a few meters of one another.

Figure 7.21. BTS on hilltops providing good signal coverage

Figure 7.22. A BTS at S36.890184^0, W73.137965^0

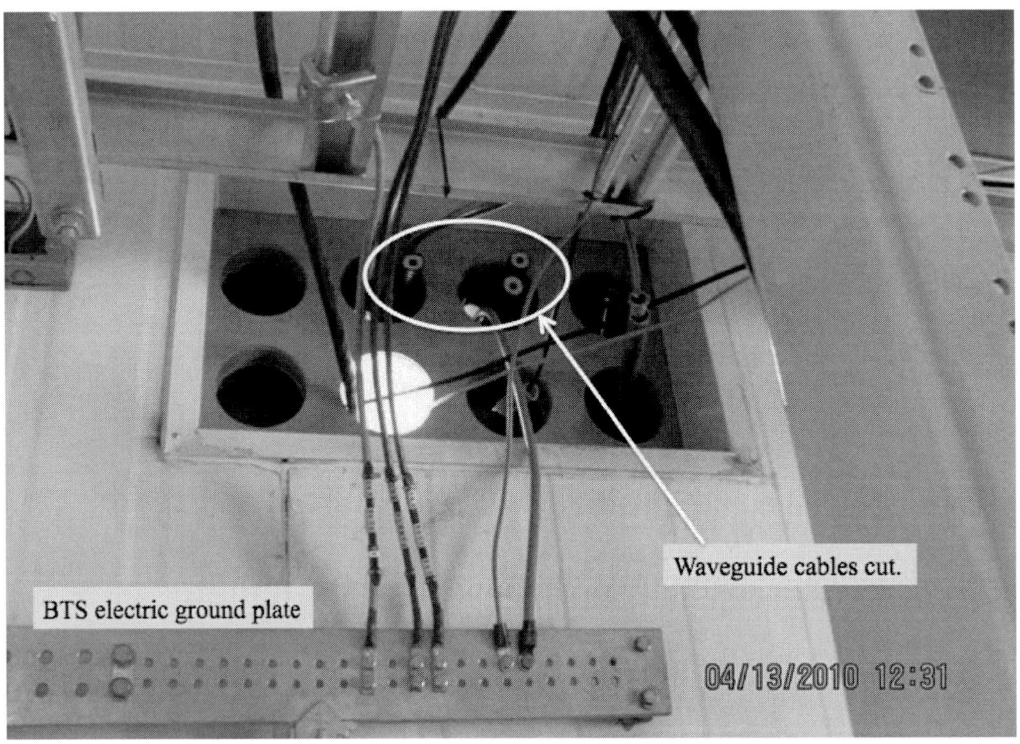

Figure 7.23. Cut waveguide cables at decommissioned site

Figure 7.24. BTS power equipment installed outside the building

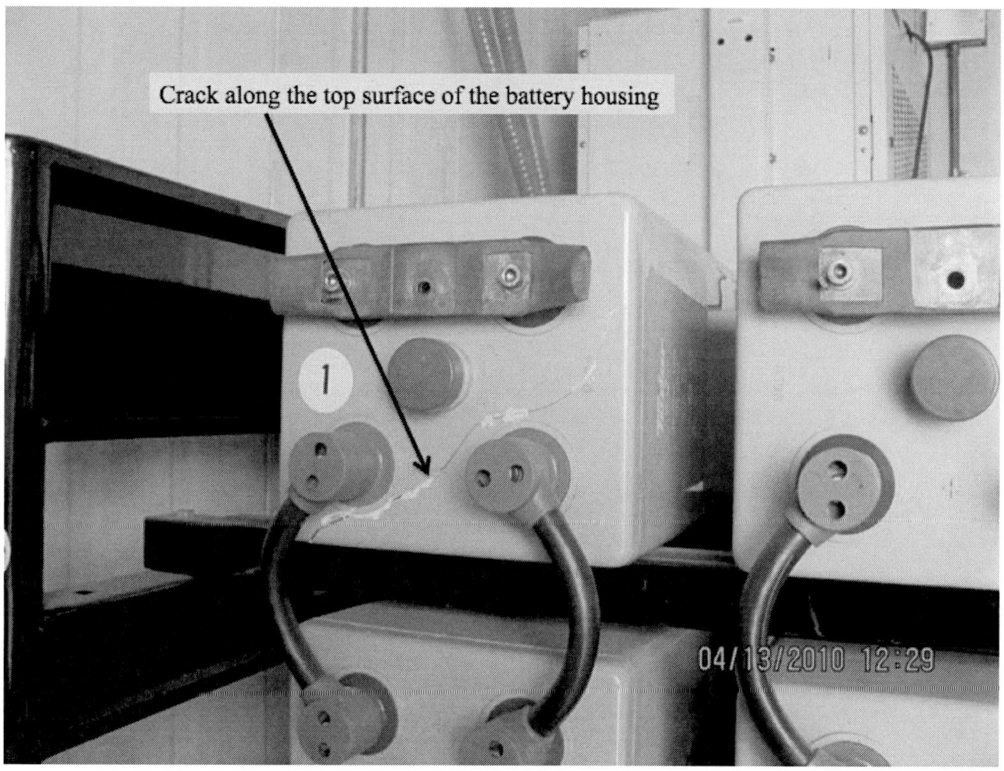

Figure 7.25. Damaged battery. Note that the batteries were not secured on the shelf.

Two antennas fell from the cable guided steel tower (Fig. 7.27). The equipment frames were anchored, but the remaining cellular equipment was not secured on the frame's shelf (Fig. 7.28).

The cable guided steel tower base did not show any signs of damage (Fig. 7.29). All towers were designed and installed based on wind load.

This site showed the typical installation practices of BTSs.

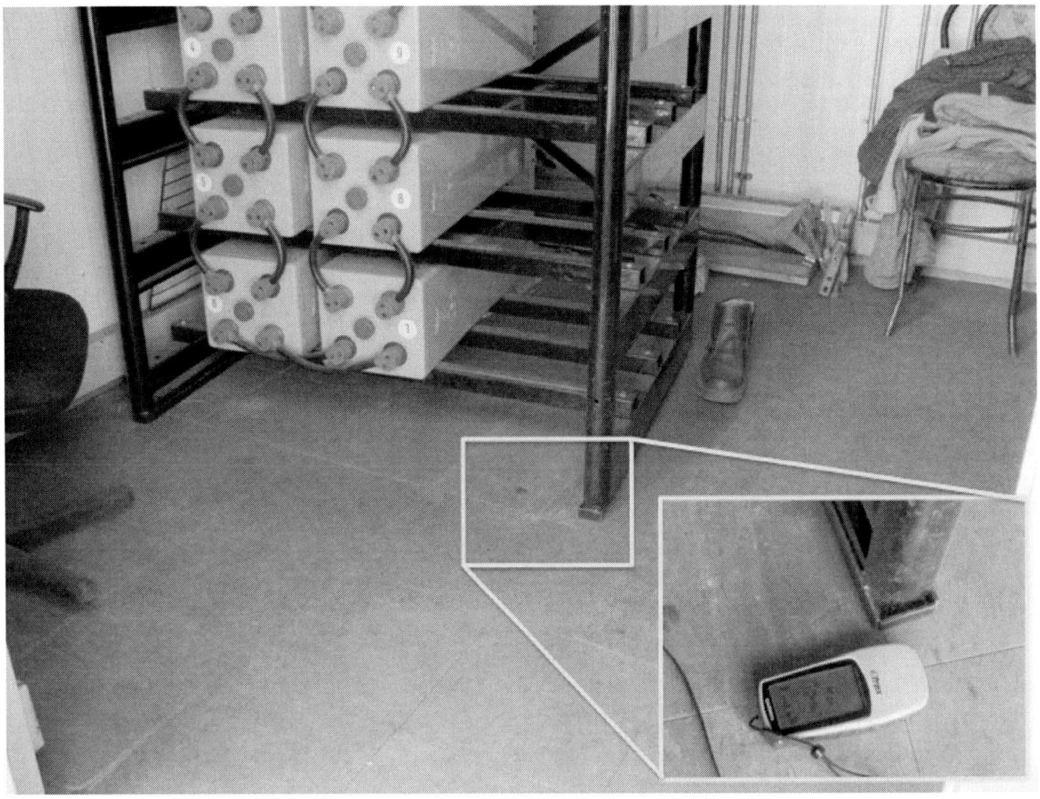

Figure 7.26. Dislocated battery rack; the damaged battery was on the top shelf

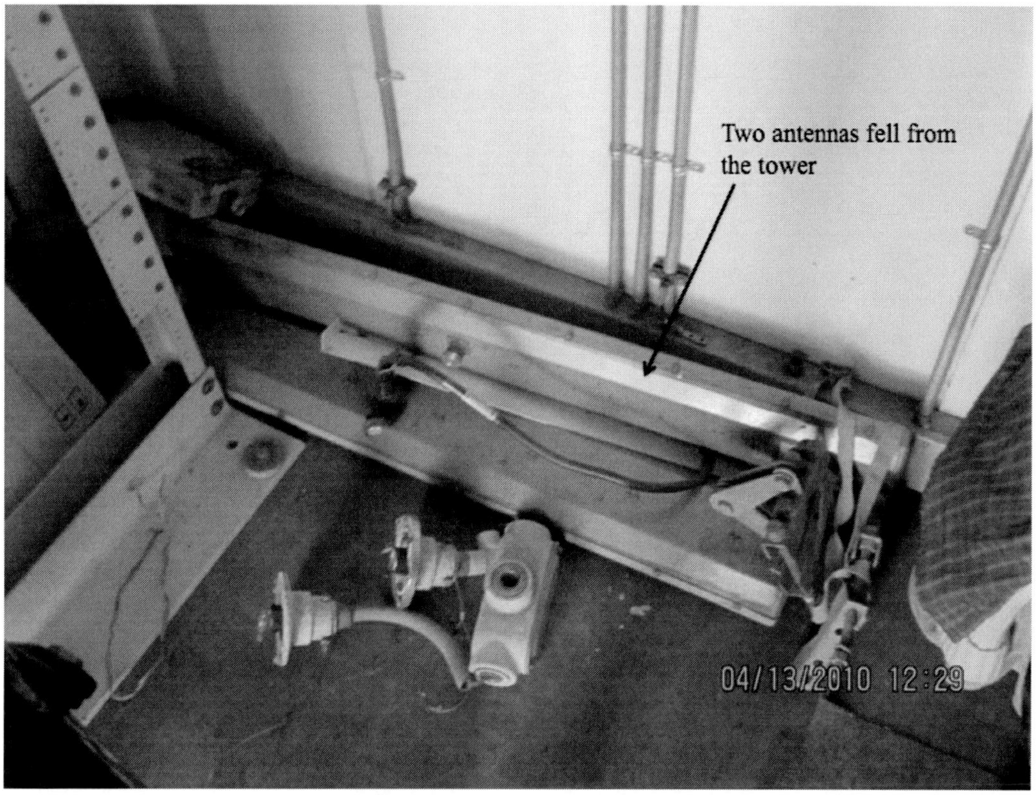

Figure 7.27. Damaged antenna

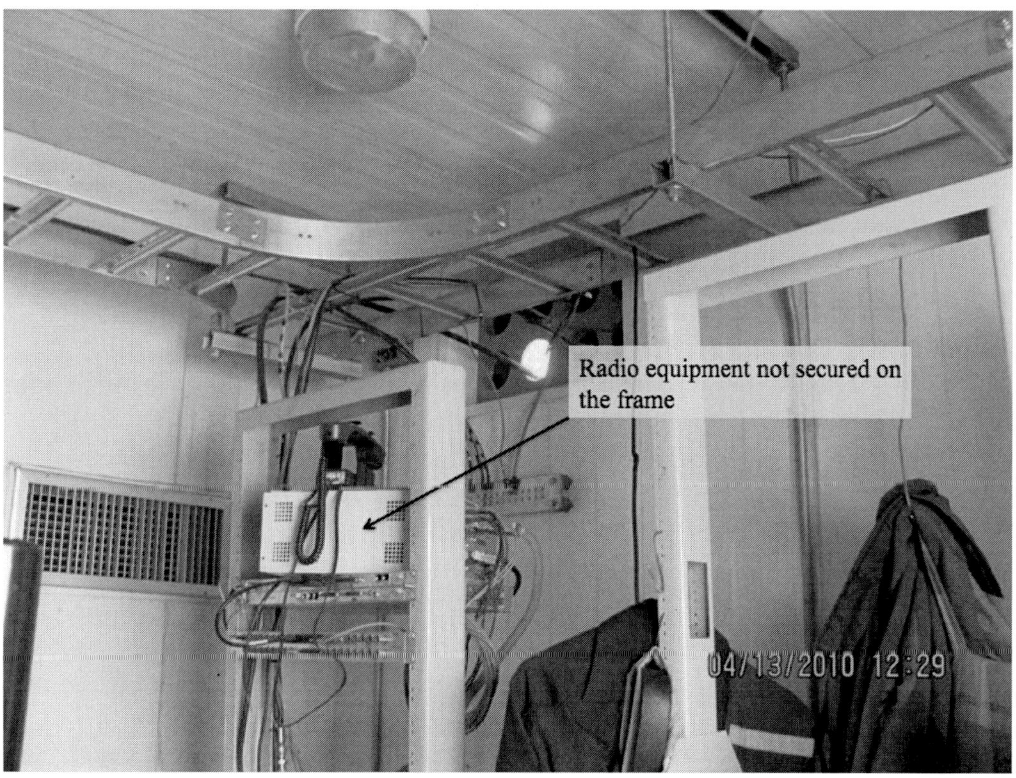

Figure 7.28. BTS equipment frames

Figure 7.29. Tower anchor showing no signs of damage

Cell Sites in Talca

Talca was one of the three cities that sustained the most damage from this earthquake. Many BTSs had small power generators installed to operate the sites before the commercial power fully recovered.

As most BTSs are unmanned and the demand on service providers' resources was high, we were not able to obtain access to all the BTSs. Many of the findings were observed from a distance outside the BTS perimeter fence. Our observations combined with information collected in meetings with service providers provide a fairly accurate assessment.

Figure 7.12 was one of the many BTSs in Talca that had a small power generator unit installed and was running 3 weeks after the earthquake.

Cell Sites in Curico, Concepción, Constitución, and Along Highway # 5

In Curico one of the many BTSs had a backup power generator unit on site (Fig. 7.30). This site is shared by two cellular service providers. Note the debris from the damaged walls on both sides. The equipment housings do not show any signs of damage.

Figure 7.30. A BTS in Curico with a backup power generator unit on site

Cell Sites Reported by Movistar

The following are the sites we received we received reports on by Movistar, a major cellular service provider in Chile, during our meeting in Santiago. Movistar is a brand name of Telefónica, originally one of the biggest telecommunication services providers in Spanish-speaking countries, which has extended its operations worldwide.

BTS Capitanía at Las Condes, Santiago (Region XIII)

This is a roof top site (Fig. 7.31). There was no damage to the building housing the equipment. There were a few equipment panels damaged (Fig. 7.32), a cracked floor (Fig. 7.33), and a cable connection pulled off the connector when the equipment fell from an equipment frame top (Fig. 7.34).

BTS Encomenderos at Providencia, Region XIII

This is another rooftop BTS site (Fig. 7.35). The anchors of one of the antenna mounting structures were pulled out, and the antenna fell on the roof (Fig. 7.36). The site inspector marked it as a critical damage due to loss of one cell.

Fortunately, the antenna and its support structure fell back on the roof instead of over the edge of the building, which could have caused some cable damage and made the repair process longer and costly.

Figure 7.31. BTS equipment building at Las Condes, Santiago (Photo reproduced with permission from its source)

Figure 7.32. Panels fallen off the equipment frames (Photo reproduced with permission from its source)

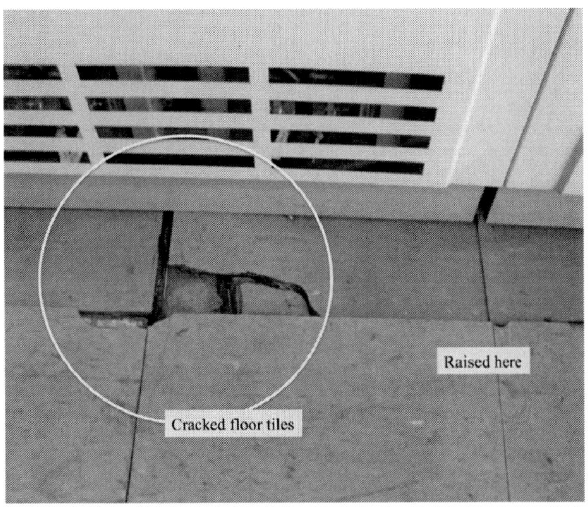

Figure 7.33. Cracked floor tiles (Photo reproduced with permission from its source)

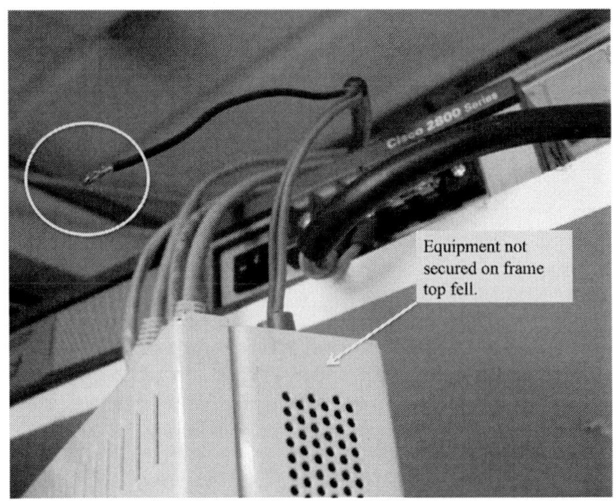

Figure 7.34. Disconnected cable and equipment hanging by the cables (Photo reproduced with permission from its source)

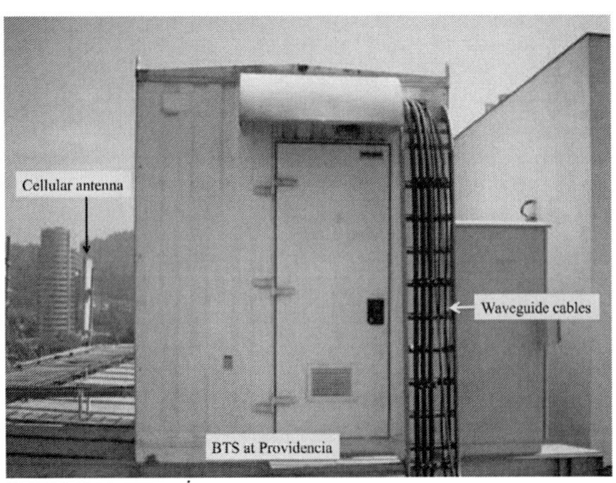

Figure 7.35. BTS at Providencia, Region XIII (Photo reproduced with permission from its source)

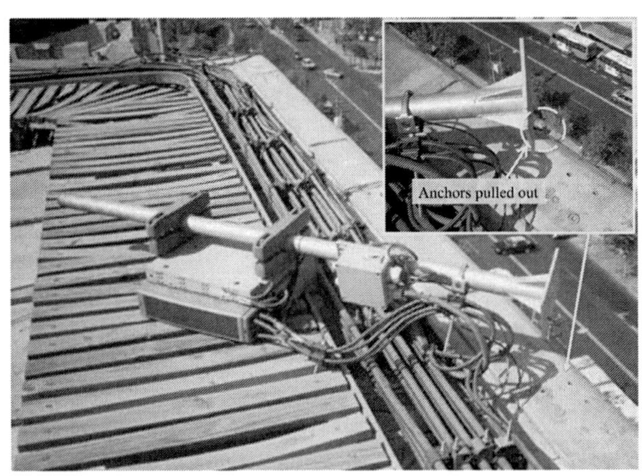

Figure 7.36. Damaged antenna (Photo reproduced with permission from its source)

BTS at SC Cañete

The antennas tower collapsed at this site (Fig. 7.37 and 7.38). The monopole was attached to the water tank, which collapsed and brought down the pole including the equipment fixed to the structure. This BTS was destroyed, including the building.

BTS Jerónimo de Alderete at Las Condes, Santiago (Region XIII)

The damaged items reported at this rooftop BTS were the battery and battery rack, transmission equipment, equipment panels, and floor. Figures 7.39 to 7.40 show the damaged equipment. This site was out of service for a few days.

Figure 7.37. Side view of the collapsed tower (Photo reproduced with permission from its source)

Figure 7.38. End view of the collapsed tower (Photo reproduced with permission from its source)

Figure 7.39. Damaged battery and battery rack (Photo reproduced with permission from its source)

Figure 7.40. Damaged transmission equipment (Photo reproduced with permission from its source)

BTS Victacura at Victacura, Santiago (Region XIII)

This BTS is in an apartment building. The air conditioning unit was damaged (Fig. 7.41); an unanchored equipment frame was dislocated, and there were cracks on walls. Figure 7.42 shows the unanchored frame dislocated from its leveling pad.

Temporary BTSs

Most of the temporary BTS set-ups were observed in Santiago. This is likely a result of damage to normal BTSs combined with the high rate of usage within the city. There were two temporary BTSs less than 1 km apart in El Golf district in Santiago.

Temporary BTS (S33.409393^0, W70.604089^0)

This temporary BTS was set up next to a small BTS. Most likely, the BTS was not functioning properly due to power outage or damage to the equipment; therefore, a mobile BTS unit was set up close by to provide backup cellular coverage to the area. Figure 7.43 shows the temporary set up.

Figure 7.41. The damaged air condition unit (Photo reproduced with permission from its source)

Figure 7.42. The dislocated frame (Photo reproduced with permission from its source)

Figure 7.43. Temporary backup to the BTS at this site and fenced in unit at the back

Temporary BTS (S33.416365^0, W70.604113^0)

This temporary BTS is situated at the corner of a T-intersection of two streets. Power was tapped from the nearby pole mounted low-voltage transformer with a power transfer box mounted on the pole. Figure 7.44 shows the BTS; Figure 7.45 shows the power feed to it.

Figure 7.44. Temporary BTS in Santiago, El Golf district

Figure 7.45. Power feed to the cellular unit

Figure 7.46. A typical telecommunication aerial cable routed along roads and streets within a city

7.4 Landline Network Including Transmission and Distribution

The landline service disruption was mainly caused by the commercial power outage. In Regions VII and VIII, the landline services for both local and long distance services were back on line about 3 days after earthquake. The massive concentration of aerial cables within major cities such as Santiago, Talca, and Concepcion can result in a large number of circuit failures and lengthy repair time after an earthquake. Figure 7.46 shows the poor aerial cable management commonly seen in cities.

Transmission is mainly achieved by using aerial and buried cables and microwave. The distribution system uses aerial cables that share the utility poles with the electric power distribution system.

Central Office in Concepción (S 36.82139°, W 73.05283°)

This small central office (CO) at the corner of Anibal Pinto and Martinez de Rozas in Concepcion also has a BTS. The tower on the southeast corner of the building is a stand-alone tower with both cellular antennas and microwave dishes for inter-CO transmission. Another cellular service provider's BTS is located across Martinez de Rozas on the north side of this CO. This BTS has a standalone tower for cellular antennas as well as the transmission network's microwave dishes.

This CO experienced a power outage and had some minor damage, but the power generator started to provide backup power. One switching unit tipped on its side and was not operating. There was also minor battery damage. The building in general performed well with a few minor cracks on walls and expansion joints.

The backup power generator unit (176 K-Watt) is installed in the courtyard just behind the cellular and transmission tower within the CO perimeter (Fig. 7.2). The unit is anchored on an I-beam frame that is bolted on the concrete pad (Fig. 7.47). It is also enclosed in a metal housing with access on both sides. A few drums of fuel for the generator were located close by (Fig. 7.48).

Figure 7.47. CO backup power generator anchoring

Figure 7.48. Extra drums of fuel for backup generator

Figure 7.49. A water tank for use during the water supply disruption

The water drum in the courtyard indicated that this location was out of water (Fig. 7.49).

Transmission System Damage

The telecommunication network's transmission system connects all the switching nodes (COs). In some case where the distance between COs is large, repeaters will be used at the appropriate locations to maintain acceptable signal quality. Therefore, the cables with repeaters must be routed close to available electric power lines.

The majority of the transmission cables are located underground in conduits or tunnels. Splicing and cross connections are placed in manholes along the conduit and tunnels. When cables have to cross rivers or ravines, bridges and overpasses are typically used to support the cables.

The cables are either copper or optical fiber. With the demand for speed and the growth of customers, more and more optical fiber cables are being installed. The disadvantage of an optical fiber cable is its high capacity; a damaged optical fiber cable affects more customers than a damaged copper cable. With the implementation of optical fiber cables, there is more reason to incorporate redundant routing to reduce service interruption.

In addition to cables, microwave is also used to connect COs and BTSs in locations where it is not economical to run cables. Microwave requires a line of sight to transmit and receive signals between sites.

Due to the spatial nature of the transmission system, which covers almost all topography of Chile, some damage is unavoidable. Proper planning of circuit routing and redundancy can render circuit capacity reduction to a minimum.

There are three modes of failure in the transmission system:

1. alecolocating optical fiber cable damage along bridges,
2. microwave dish alignment, and
3. power outage.

Colocating

With high incidents of bridge damage and collapse, the chances of severed telecommunication links become high. In Concepcion, it seems unusual that only one out of three bridges that cross Biobio River have a telecommunication link colocated with the bridge. It was also unusual that all three bridges sustained damage with collapsed sections. The optical fiber cable that was supported along the bridge was severed due to a collapsed section at the abutment. Figure 7.8 shows the damaged cable and the splice box under the bridge.

Microwave

Most of BTSs in Regions VII and VIII use microwave as their interconnection with the MTCO, CO, and other BTSs (Fig. 7.50). Microwave interconnection depends on line-of-sight; therefore, over a long distance a small angle of rotation will sever the connection. The microwave dishes are secured to the tower using metal clamps similar to pipe clamps. Because these depend on friction to secure the assembly to the tower, they are easily loosened under strong vibration unless there is positive locking device such as a pin or a screw that prevent the assembly from rotating. Figure 7.51 shows a typical assembly securing the microwave dish to the tower.

Figure 7.50. Extensive use of microwave in a telecommunication network

Figure 7.51. U-clamps secure the microwave dishes

Power Outage

Coupled with the poor power management of many telecommunication facilities, power outage was the main cause of telecommunication disruption in this earthquake. Power outage in the transmission network was not significant, as most COs were installed with power generator units. Only a handful of COs had generator problems.

Distribution System (Outside Plant) Damage

Power outages also affected active outside plant network elements, such as Digital Loop Carrier (DLC) RTs (Fig. 7.50). As the link between these RTs and their host at the CO is usually made with fiber optics, RTs need to be powered locally. As shown in Fig. 7.52, DLC RTs very rarely have fixed permanent generators, so their only power backup comes from batteries, usually providing up to 8 hours of autonomy. Although in the United States service to DLC RTs is typically restored with small portable generators, in Chile service was restored only after power from the grid was back on. Still, the number of DLC systems in Chile is limited so their failure affected a relatively small number of subscribers.

Figure 7.52. Most DLCs are installed on poles with power supply coming from a nearby distribution system. Reserved power is very limited.

The main cause of damage to this system was collapsed houses and buildings and utility pole damage severing the drop wires and local cable routes. However, the impact was quite localized, particularly in Talca where many URM buildings failed. While there is no need for a phone line in a collapsed house, the severed cables can affect undamaged houses nearby as telephone services are critical immediately following an earthquake.

Poor cable management of the distribution system as observed in the aerial cable networks is one of the reasons of the slow recovery (Fig. 7.53).

Figures 7.54 to 7.61 highlight the distribution system damage in regions VII and VIII. These photographs were taken 3 to 5 weeks after the earthquake. Restoring service is a high priority in emergency recovery. Rebuilding to restore normal conditions, including cleaning up the debris, is a lengthy process.

Figure 7.53. Transmission and distribution aerial cable congestion

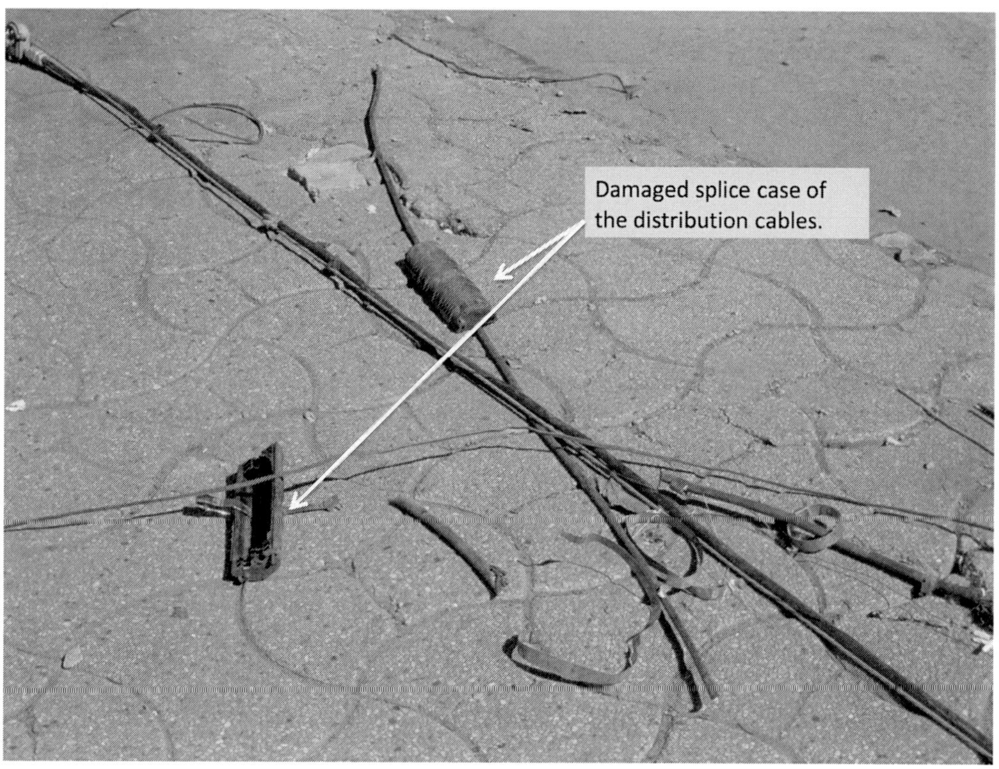

Figure 7.54. Damaged distribution cable splice case

Figure 7.55. Aerial cable damage, showing the problem of poor cable management

Figure 7.56. Distribution cable damaged by building collapses

Figure 7.57. Drop wire severed due to building collapse

Figure 7.58. Distribution cable damage due to pole collapse

Figure 7.59. Drop wires severed by fallen debris

Figure 7.60. Drop wire severed; partial building collapse

Figure 7.61. Utility pole damage in Pelluhue

7.5 General Observations

Two sites in Concepción were visited during the investigation—a cellular base station (aka BTS) and a central office. We also met with two major service providers in Chile to collect general system performance information. One meeting was held in Concepción, the other in Santiago. This section provides our observations at these two sites and the observations of the service providers from the meetings.

Our overall impression of the Concepcion CO was fair; it is a small CO within the region. The number of damaged outside plant cables was very low. The damage to communication equipment and battery banks inside service provider facilities was more than expected. In most cases this resulted from inadequate anchoring and securing of components on equipment frames and shelves.

The backup power generator unit at the Concepcion CO functioned well in the earthquake. It was unclear whether the barrels of fuel were brought in after the earthquake or were normally available at this site.

Both antenna and tower performance in the cellular system were surprisingly poor, contributing to the longer recovery time.

Many BTSs installed on the roofs of commercial buildings were disconnected from the network because the buildings were condemned. Regardless of whether there was any damage to the equipment installed onsite, the power supply to these buildings was disconnected. In some cases we observed that the antennas were damaged.

7.6 Major Observations and Recommendations

The dry season, which reduced the number of ground failures and landslides, might have contributed to the low incidence of cable failure—both aerial and underground. The three communities (Isla Rocuant at Talcahuano, Dichato north of Concepción, and Pelluhue south of Constitución) visited by ASCE/TCLEE team members that were hit by the tsunami showed complete destruction of the aerial cable systems.

For cable conduits colocating with bridges, a more flexible connection to the bridge should be used to avoid damage from the displacement of the bridge deck. One option is to place more cable slacks inside the conduit to allow for displacement, thus using a larger diameter conduit with long cables curling inside the conduit to allow for displacement of severed conduit by the bridge deck. However, this would not prevent cable cut due to bridge deck collapse. The other options for crossing a river are to (1) build a dedicated structure to secure the cable conduits, and (2) use redundant cables or microwave.

We support the service providers' conclusion that they need to improve their backup power capability. For remote or difficult to access sites, longer battery reserve power should be considered. We discussed the options of using solar and wind power for these sites, but in general, all telecommunication service providers should increase battery power reserve and install backup power generators. A maintenance policy must be implemented along with the backup power generator installation to ensure reliability at the time of need. Good practices for power generator installation are readily available and can be implemented inexpensively.

Equipment frames should be adequately anchored. The components on shelves should be secured by either bolting or strapping them to the shelves. This is the lowest cost insurance to prevent earthquake damage. Battery racks should be adequately anchored, and the batteries on the shelves should be secured to the rack to prevent them from moving around and damaging each other when the earthquake hits.

More Cellular on Wheels (COWs) should be available for emergency response (Fig. 7. 63). These units can be deployed quickly to maintain coverage.

Finally, a separate radio system should be implemented for emergency response, particularly for the service department as the primary dispatch tool.

Corrective Actions (received from Movistar in February 2011)

Government of Chile

As a post-earthquake effort to reduce future problems during an earthquake, the government of Chile, through its Under-Secretary of Telecommunications, is requiring all network operators to be able to operate a minimum network under any condition for at least 48 hours. An early warning message sent via Short Message Service (SMS) is being implemented for the cell phones within the hazard area. Short messaging will also be promoted during catastrophes to avoid network congestion.

Figure 7.62. A mobile unit installed with equipment to temporarily replace an impaired BTS is called Cells On Wheels (COW). Notice the use of the satellite link to connect to its host switching center. This COW was located a block away from the cell site shown in Figure 7.13.

Movistar (Service Provider)

Movistar has retrofitted all DLC remote cabinets with a power converter so that they can be powered from the central office via copper pairs. This reduces the risk of power outage experienced during this earthquake.

7.7 Acknowledgements

The authors are grateful to Ms. Francisca Vilches, Mr. Victor Morales, and Mr. Luis Leal of ENTEL, Concepción, as well as Mr. Carlos Urzua and Mr. Jorge Martinez of Telefonica, Santiago, who all provided valuable information and photos for this monograph. We are also indebted to Ms. Sonia Castellanos of Emerson Network Power, who assisted us in coordinating the meetings and provided us with the network performance information after the earthquake. Professor Alexis Kwasinski's translations during our meetings and his contributions with edits and photographic material to this chapter are much appreciated. We are sure that these contributions will help to improve the performance of telecommunication networks in future earthquakes.

Unless otherwise noted, all photos were taken by the ASCE/TCLEE team members.

8
WATER AND WASTEWATER SYSTEM

The Chile M_w 8.8 subduction zone earthquake on the Nazca Plate struck central and southern Chile on February 27, 2010 at 3:40 a.m. (local time). The fault rupture area was approximately 500 km (north to south) by 100 km (west to east). The northernmost extent of fault rupture was near the city of Rancagua, or about 50 km south of Santiago. The southernmost extent of fault rupture was near the city of Temuco, about 550 km south of Santiago.

Ground Motions

Chile is divided into 15 regions, commonly referred to using Roman numerals I through XV, plus the capital region of Santiago.

The regions most affected by the earthquake (either directly over or directly east of the ruptured plate) were Region VI (northern extent of fault rupture, including the cities of Rancagua, San Fernando, and Talca), Region VII (including the cities of Curico, Talca, and Constitución), Region VIII (including the cities of Concepcion, Chillan, Los Angeles), and Region IX (southern extent of fault rupture, including the city of Temuco). Along the coast of the Pacific Ocean is the city of Concepcion, which has a current population of about 1 million, and the mid-sized cities or towns of Constitución, Talcahuano, Coronel, Lota, and Lebu.

About 4.5 million people live directly over or just east of the rupture zone, the bulk of whom experienced horizontal ground motions between PGA = 0.15 g (firm soil conditions at the east foothills of the Andes) and 0.40 g (firm soil conditions at the Pacific Ocean). Local variations in ground motion were about -50 percent to +100 percent of these values.

The University of Chile maintained a set of strong ground motion instruments. The uncorrected PGA maxima are listed in Table 8.1 and the corrected (bandpass filtered) velocities in Table 8.2 (Boroschek et al. 2010). The highest two recordings were on rock sites (0.6 g, 0.9 g). Above the rupture zone, horizontal PGAs at firm soil sites were commonly 0.25 g to 0.45 g. The University de Chile recording at Concepcion contains a lot of long-period motion (T = 1.5 to 2.2 seconds), suggesting a soft soil site. Most of the other instruments reflect firm soil conditions.

Immediately north of the most strongly-shaken regions are Region V (Valparaiso along the coast) and Chile's capital, Santiago, towards the east. The population of Santiago is about 5.5 million people. These regions experienced ground motions typically between PGA 0.20g to 0.30g, but again locally with variations of -50 percent to +100 percent or more of these values.

Table 8.1 Peak Ground Accelerations (g) (Uncorrected)

Station	Region	Azi-muth	PGA NS g	PGA EW g	PGA Long g	PGA Trans g	PGA Vertical g
Copiapo	III	0	0.030	0.016			0.008
Vallenar	III	0	0.020	0.019			0.010
Papudo	V	60			0.295	0.421	0.155
Vina del Mar Marga-marga	V	0	0.351	0.338			0.261
Vina del Mar Center	V	0	0.219	0.334			0.186
Valparaiso, UTSFM	V	180			0.224	0.266	0.146
Valpariaso, Almendral	V	310			0.137	0.304	0.080
Llolleo	V	340			0.319	0.564	0.702
FCFM	Santiago	0	0.165	0.163			0.138
Center	Santiago	270			0.218	0.309	0.182
Maipu	Santiago	0	0.561	0.478			0.240
Penalolen	Santiago	0	0.295	0.293			0.280
Puente Alto	Santiago	0	0.265	0.263			0.130
La Florida	Santiago	0	0.236	0.165			0.130
Matanzas	VI	0			0.342	0.308	0.234
Hualane	VII	0			0.389	0.461	0.390
Curico	VII	150			0.470	0.409	0.198
Talca	VII	0			0.477	0.424	0.244
Constitucion	VIII	0			0.552	0.640	0.352
Concepcion U.de.Chile soft soil site	VIII	60			0.402	0.284	0.398
Concepcion U de Concepcion Rock site	VIII						
Angol	IX	0	0.928	0.681			0.281
Valdivia	XV	0	0.092	0.138			0.051

Table 8.2. Peak Ground Velocity (cm/sec) (Corrected)

Station	Region	Azi-muth	PGV NS Long cm/sec	PGV EW Trans cm/sec	PGV Vertical cm/sec	Spectral Dominant Hz Period (Seconds)
Copiapo	III	0				0.3 - 0.4
Vallenar	III	0				0.25 – 0.95
Papudo	V	60	16.7	24.8	6.1	0.3
Vina del Mar Marga-marga	V	0	37.9	44.6	12.2	0.35 - 0.85
Vina del Mar Center	V	0	20.9	32.6	13.3	0.5 – 0.8
Valparaiso, UTFM soft	V	180	7.4	16.0	9.4	0.2
Valpariaso, Almendral	V	310	29.2	22.3	10.7	0.7 – 0.9
Llolleo	V	340	25.8	31.0	22.3	0.25
FCFM Lee	Santiago	0	15.6	22.0	15.7	
Center	Santiago	270	21.9	25.6	14.4	
Maipu	Santiago	0	44.1	39.0	21.8	
Penalolen Hospital Oriente	Santiago	0	29.3	22.7	15.3	
Puente Alto Hospital Sotero del Rio	Santiago	0	24.6	31.5	16.2	
La Florida Mirador Station Metro	Santiago	0	19.2	17.9	12.4	
Matanzas	VI	0	43.4	27.8	15.7	0.5
Hualane	VII	0	38.8	35.0	26.2	
Curico	VII	150				0.2
Talca	VII	0	27.4	33.4	19.1	
Constitucion	VIII	0	43.3	68.6	20.5	
Concepcion soft	VIII	60	67.3	51.7	26.7	1.5 – 2.2
Angol rock	IX	0	34.3	37.6	12.2	0.2
Valdivia	XV	0	13.6	18.4	6.6	

Geotechnical Considerations

In addition to the effects of ground shaking, buried water system pipelines are strongly affected by local geotechnical effects, including liquefaction, landslide, and surface faulting. We observed no evidence of surface faulting; this is as expected, as there was no land-based surface rupture in this event. While there was evidence of deep-seated landslides in rural areas, we observed no evidence of landslides in major urban areas. The topography of Chile is similar to California, with coastal hills at the Pacific Ocean, a flat central valley, and high mountains to the east. There were few landslides observed in the strongly shaken coastal hills, and none in the Andes. The paucity of landslides might be attributed, in part, to the time of year of the earthquake (February 27), which corresponds to the end of the dry season. Had the earthquake occurred in the wet winter season (say, August), more landslides may have been triggered.

There was widespread evidence of liquefaction along the coastline of the Pacific Ocean as well as along the banks of the rivers that empty into it. Liquefaction effects were particularly common in Concepcion and within 50 km north and 150 km south along the ocean. Perhaps 30 percent of riverbanks in these areas showed clear movement from a few inches to a few feet laterally, with some vertical settlements. The geology along these areas is commonly alluvial, with sands and silts. At 14 ports visited, nearly all sea walls showed evidence of rotation (a few did not), with corresponding damage to piers and wharves as well as buried infrastructure immediately on the landward side of these facilities. Bridge abutments at the sides of rivers were commonly tilted or otherwise damaged where there was evidence of liquefaction nearby. A few bridge piers within rivers showed evidence of settlement (less than 5 percent of all piers). Where liquefaction occurred, there was severe damage to buried water pipes and other buried infrastructure, as discussed in following sections.

Subsidence and land uplift also occurred. This effect is due to the rebound of the South American Plate as it is released along the interplate zone. Areas in the western side of the rupture zone uplifted; areas to the east subsided. This type of ground movement is not thought to be of general concern to buried pipelines, and the ground curvatures are not high enough to materially stress buried pipe. However, these uplifts and subsidences affect the ports and harbor areas, as they change the sea level and affect the ability of shipping traffic to pass through shipping channels.

This earthquake also caused tsunamis. The tsunamis impacted the city of Constitución and an island in the river adjacent to Constitución, resulting in many fatalities in those areas. Tsunamis also inundated the port of Talcahuano, resulting in widespread damage to fish factories and port facilities in that important commercial area. Tsunamis are not known to have directly inundated any water treatment, wastewater treatment, or buried water/wastewater pipes.

8.1 Description of System

Water Systems

Two water companies deliver potable water to the majority of the cities in the strong shaking area. Aguas Andina serves the Santiago region. Essbió operates individual water systems serving the cities of Rancagua, Curico, Talca, Chillan, Concepcion, Talcahuano, Coronel, Los Angeles, and Temuco. In addition to these two large companies, there are several hundred small rural water systems developed by the government's Ministry of Public Works (MOP). MOP also maintains many canals throughout Chile. The following summarizes the major performance issues for these water systems.

Essbió Potable Water

Essbió serves potable water in urban communities, totaling about 4 million people served. Each community's system is separate and not interconnected. In total, the water systems include about 7,000 km of transmission and distribution pipe, of which 3,500 km are in Region VIII and 1,200 km are in the city of Concepcion.

By far the largest amount of damage to the various Essbió water systems was concentrated in Concepcion and the adjacent community of Talcahauno. Figure 8.1 shows a plot of the restoration of potable water service for these cities. This plot was developed in conjunction Essbió, and reflects a preliminary assessment. Water was restored to more than 99 percent of customers in the Concepcion region by the first week of April 2010.

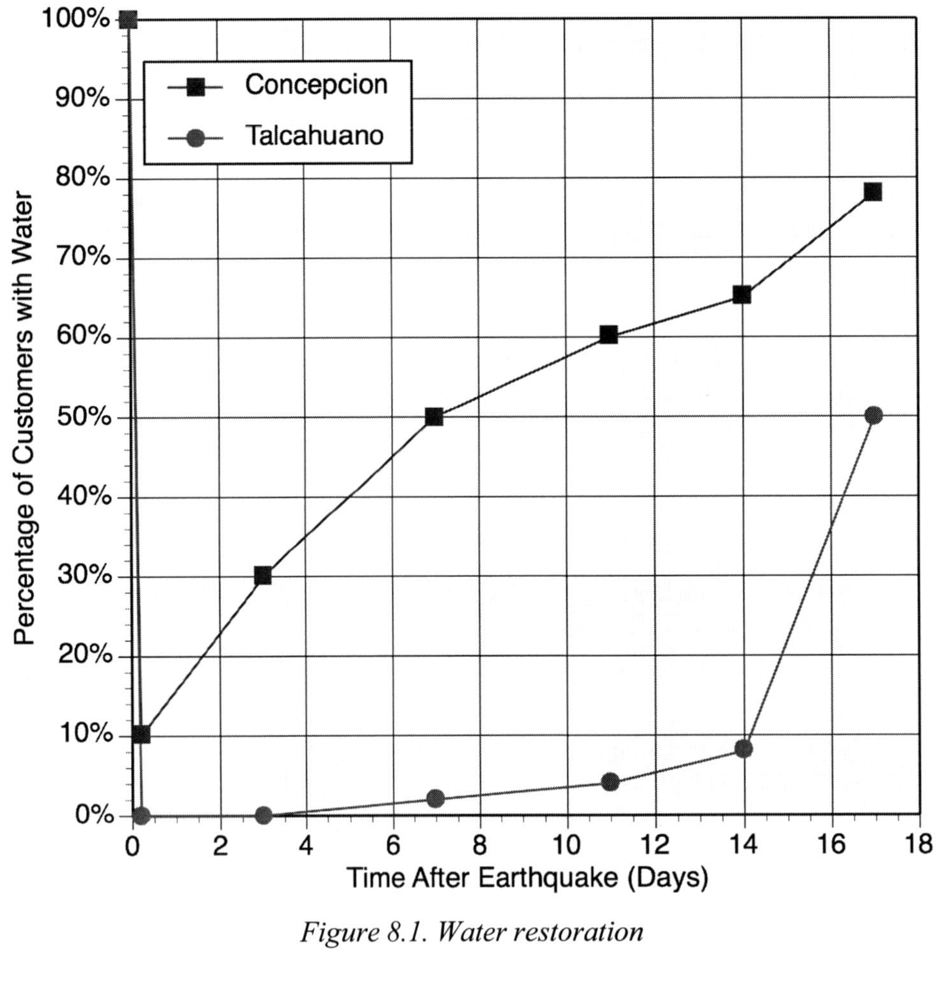

Figure 8.1. Water restoration

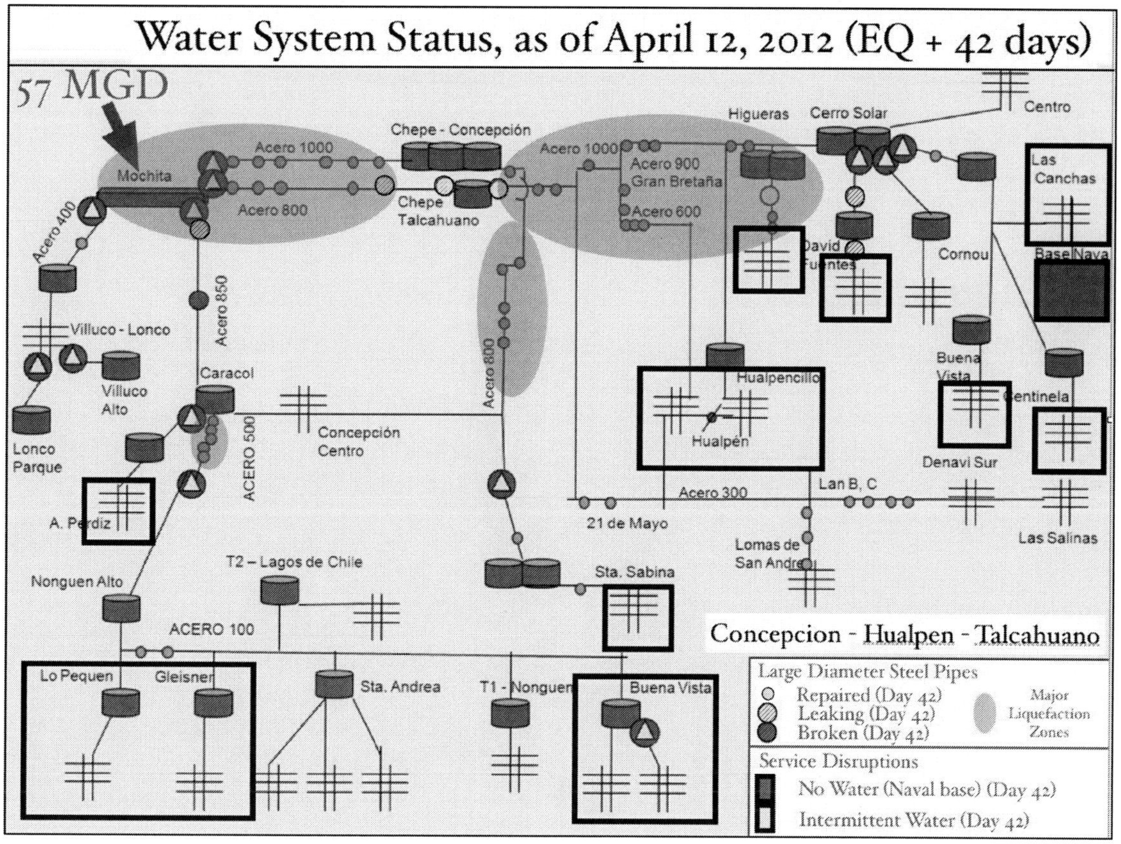

Figure 8.2. Concepcion and Talcahuano water systems

Figure 8.2 shows a schematic diagram of the potable water system serving Concepcion and Talcahuano.

The Concepcion water system includes one water treatment plant (Mochita WTP, estimated PGA = 0.30g to 0.40g), a large pressure zone serving the downtown and nearby low-elevation areas of Concepcion, and a variety of small pressure zones serving hilly terrain. Figures 8.2 and 8.3 show the major facilities and pipelines in the water system. In both Figures 8.2 and 8.3, darker gray zones indicate areas that experienced liquefaction with corresponding damage to large diameter (16 in. to 40 in. diameter) pipes. The small circles in Figure 8.2 show the schematic locations of individual breaks of larger-diameter welded steel water pipes. The solid gray zone in Figure 8.2 schematically shows the continuing water outage at the Naval base just north of Talcahuano. Note that in Figure 8.2, "Acero" means welded steel pipe, and diameters are listed in millimeters.

Figures 8.2 and 8.3 show that every major transmission pipeline from the Machito WTP to the downtown area (heading easterly or northerly) suffered damage immediately after the earthquake. A smaller residential area east of the WTP was able to get water as soon as the emergency generator at the WTP was activated, about 45 minutes after the earthquake.

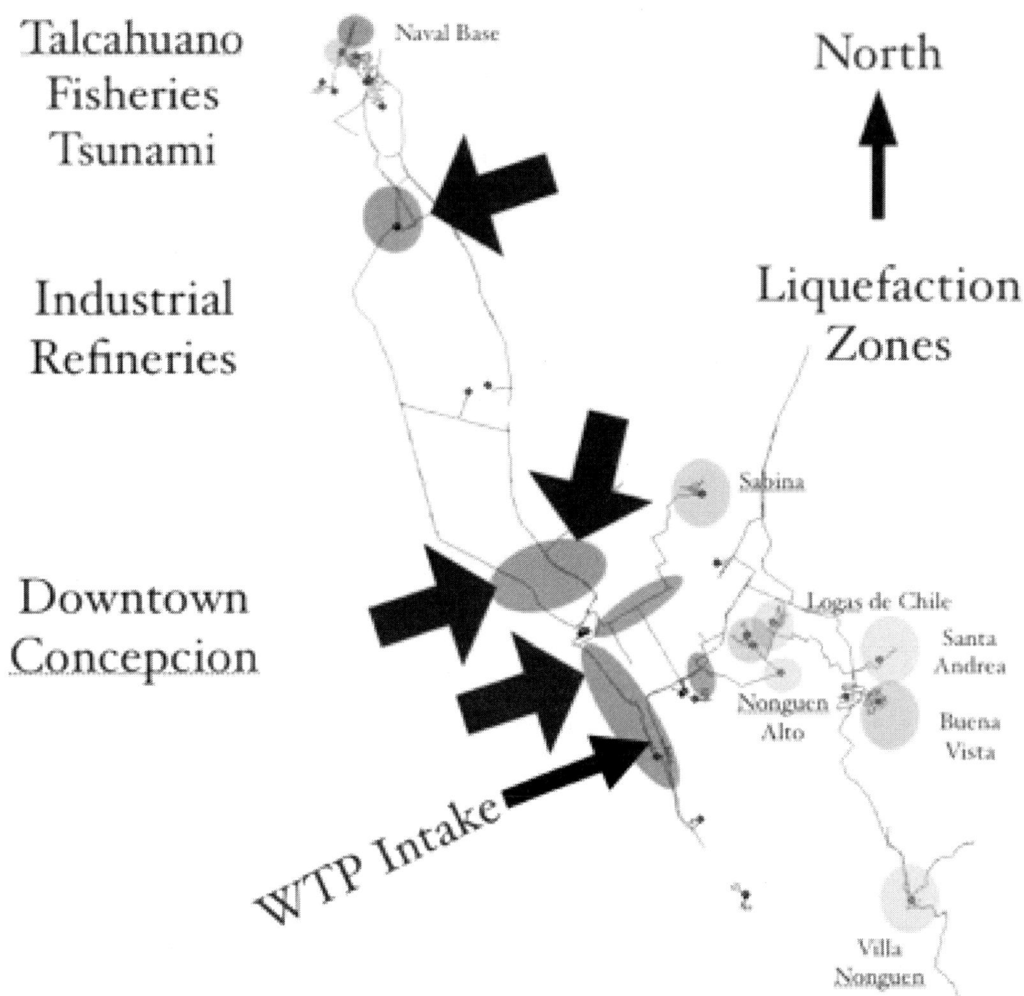

Figure 8.3. Major pipes and liquefaction zones in Concepcion and Talcahuano

The WTP has a raw water pump station that takes surface water from the Bio Bio River, treats it via four clarifiers (settling, flocculation), filters it, and then pumps it into the main pressure zone. Plant capacity is 2.5 m^3 per second (57 million gallons per day, MGD). Disinfection is by chlorination.

The raw water pump station includes a reinforced concrete structure supporting trash racks and a reinforced concrete wet well that is below the pump station floor and is housed in a reinforced concrete frame structure with some infill unreinforced brick panels (Fig. 8.4). The trash rack structure underwent about 1 ft. of lateral spread and 1 in. of settlement due to liquefaction at the shore of the river inlet. The wet well and pump station structure above did not move permanently, resulting in a 13-in. wide space between the buildings. The pump station building suffered extensive damage but did not collapse. To protect the pumps, steel "catch" structures were built after the earthquake, and the pump station building was shored up and laterally braced.

Water leaves the pump station to one of four circular clarifiers. For all four clarifiers, the center post, each set of baffle rings, and many tube settlers were damaged or dislodged by sloshing/water loads (Fig. 8.5). One of the four clarifiers was still being repaired 6 weeks after the earthquake, while the three others remained in service (albeit with short-circuits and lower efficiency).

After the clarifiers, the water goes to the 16 filters. None of the filters suffered permanent damage. All pipes in the filter pipe gallery remained leak-tight and undamaged.

All six desktop monitors in the control room fell; two broke. All six computer towers also fell from the desk to the floor. Fortunately, all six still worked. About one-third of all glassware and countertop components in the water quality laboratory fell to the ground as well; most of them broke. About half the desktop equipment in the water quality lab slid off countertops and fell to the floor, with most of them breaking. In addition, a suspended ceiling fell, and underground large-diameter water pipes within the plant boundary broke.

The treated water transmission lines leaving the WTP are welded steel, commonly 500 mm to 1000 mm diameter. There were 72 failures in these pipes. Figure 8.23 shows a wrinkled steel pipe with a tear and leaking water. Most of the failures appear to be attributed to settlements and lateral spreads. Figures 8.6 and 8.24 to 8.29 provide additional examples of damage to large-diameter transmission pipes.

The distribution system pipes (commonly 150 mm to 300 mm) include cast iron pipe (used in the oldest areas in Concepcion), asbestos cement (a material commonly used during the growth of the community), PVC, or HDPE. Today, all new distribution pipes are either PVC or HDPE. The Concepcion water system includes about 1200 km of pipe. Through April 12, 2010, repairs were made at about 3000 locations, of which about one-third were to pipe mains and two-thirds were to service laterals, which run up to the customer's meter. Figures 8.33 to 8.36 provide typical examples of repairs to distribution pipes.

Before the earthquake, the net leak rate (lost water) in the system was about 40 percent. As of April 12, 2010, it was about 60 percent. On February 27, 2010 (after the earthquake), 17 percent of Essbió's customers in Region VIII (including Concepcion, Talcahuano, and other cities) had piped potable water. By April 12, 2010, this had increased to 97 percent. Factors contributing to the water outages and restoration time were broken pipes, mobilization of a large pipe repair crew, and nearly complete loss of cellular phone service for several days (Essbió had no private radio system). Loss of offsite power was problematic, but the WTP's diesel generator was turned on 45 minutes after the earthquake and was sized to operate the entire plant at 50 percent capacity. There were also failures of a few potable water tanks in the distribution system as well as outages, as pump stations to small pressure zones did not have permanent generator sets.

Preliminary evaluations show that the bulk of pipe damage was due to permanent ground deformations (PGDs) induced by liquefaction. Recently installed fusion-butt welded HDPE pipe suffered no pipe failures, demonstrating that this type of pipe can be very effective for accommodating substation ground failures.

On February 27, 2010, Essbió was able to provide piped potable water service for 40 percent of its customers in Region VI, 25 percent in Region VII, and 17 percent in Region VIII, with the largest outages in Concepcion. By April 10, 2010, Essbió was providing piped water service for 100 percent of its customers in Region VI, 100 percent in Region VII, and 99.5 percent in Region VIII.

The last area in Region VIII to have its potable water service restored will likely be the small community of Arauco, with 2000 customers. This seaside town was devastated by the earthquake, with PGDs damaging the water infrastructure as well as the other built environment.

Rural Potable Water Systems

During the past 50 years, Chile's federal government of has constructed nearly 2000 small rural potable water systems country-wide, of which about 420 were in the strong earthquake-shaking areas. The common system uses a well with a submersible pump (about 100 to 600 ft. deep, depending on the local aquifer), an elevated water tank (commonly a steel tank atop steel braced structure, with tank storage volume of about 10,000 gallons), and poly-vinyl chloride (PVC) distribution pipe. In Regions VI, VII, and VIII, at least 73 of the elevated tanks completely collapsed. Tank failures were due to inertial overloads.

The three most common styles of elevated tanks are steel tanks on steel cross-braced structures (very common), concrete tanks on concrete cylindrical support structures (somewhat common), and concrete tanks on concrete frame structures (least common). The larger volume tanks (up to about 500,000 gallons) were all concrete. As previously noted, the elevated steel tanks on steel structures suffered many collapses. We observed dozens of elevated concrete tanks; only one, which was on a concrete frame-support structure, had collapsed.

Figure 8.7 shows a common elevated steel tank, which was undamaged. It is estimated that there are more than 1000 nearly identical tanks situation throughout rural Chile. Figure 8.8 shows a steel tank resting on a steel beam. The tank is welded, but the welds are not shown in the photograph. The figure also shows one of the four tubular column-upper frame connections; it is suspected that this bolted beam-to-column connection was the weakest link that led to the many collapses. Figure 8.9 shows one of the collapsed tanks. It was reported that in a few cases tanks damaged nearby structures as they collapsed. We obtained conflicting data about the total number of tanks that collapsed. We verified 73, but one official indicated there were more than 400 in total. We estimate that more than 75 percent of these elevated steel tanks collapsed if they experienced ground motions of PGA much over 0.25 g.

Water service restoration in the rural areas was conducted as follows: First systems with collapsed tanks were identified, and regional power was restored to the well. Collapsed tanks were then isolated from the water system; and for systems with damaged tanks, the well was operated in direct mode to the distribution system. Before the earthquake, essentially none of the more than 1000 rural water system had emergency generators.

Santiago Potable Water System

The Santiago water system includes two large water treatment plants on the east side of the city (PGA perhaps 0.10 g). The bulk of Santiago is built on firm soils or thin soils over rock, and PGAs ranged from about 0.10 g (eastern hilly areas) to 0.35 g (southwestern areas); locally ±50 percent of these values. Potable water outages in the Santiago area were reported to be sporadic and were not concentrated in any specific geographic area. Many areas had no outages. While there was widespread evidence of minor to moderate damage to many buildings in Santiago as well as some building collapses, there was no widespread evidence of liquefaction in the Santiago area. The lack of liquefaction is the primary reason for the relatively good performance of the water system in the area.

Santiago Airport Water Tank

There are four large liquid fuels and one water tank located at one site the Santiago Airport. All five tanks are welded steel. The tank site is located in Chile's Seismic Zone 2, which translates to a seismic design level motion of PGA = 0.30 g.

The water tank collapsed; however, the adjacent four liquid fuels tanks remained intact (Fig. 8.10, 8.11, and 8.12). The water tank had a storage capacity of 1300 m^3, or about 340,000 gallons. The water tank was about 50 ft. in diameter, was about 23.5 ft. high to the overflow level, and had a lower steel course wall thickness of 5 mm. Computed internal hoop stress due to water pressure is about 16,000 psi, which is within the typical range for allowable stresses for common design of steel water tanks.

The concrete tank was unanchored and rested on a concrete at-grade ring beam. Water pipes were attached to the tank at the lower course, and the steel roof was supported by beams. There was ample evidence of internal corrosion to the steel at and near the roof level; however, the outside of the tank was painted and did not appear to have much corrosion.

The observed failure mode appeared to be tearing of the bottom course from the steel floor plate, with a nearly uniform tear vertically along one of the vertical seam welds in the lower courses (Fig. 8.12). This led to collapse of the tank, with subsequent buckling and tearing of the steel.

The uplifted floor plate seen in Figure 8.11 (about 3 in. of permanent vertical uplift) strongly indicates that tank wall uplift occurred during the earthquake. For unanchored at-grade steel tanks, this is expected performance.

We would not normally expect this type of performance (gross collapse) for a well-built and well-maintained steel tank in a magnitude 7 earthquake. The possible root cause(s) for the observed failure are as follows:

- Tank wall uplift leading to tearing of the bottom course at the floor plate welds. This is the most likely cause of failure. This clearly demonstrates the fallacy of assigning R = 3.5 to 4.5 for unanchored steel tanks (implied by all AWWA D100 code versions since the early 1970s) coupled with the long duration of ground shaking.

- Accumulating corrosion could have weakened welds and steel plates. We observed a modest amount of corrosion on the exterior of the tank. The interior of the tank appeared to have a fair amount of corrosion below the water line and major corrosion above the water line.

- Once the tank wall-floor plate tore due to a combination of wall uplift (repeated cycles) and wall buckling, the tear likely propagated up a vertical full penetration weld. With the tearing of the tank wall, the high rate of release of water through the vertical tear would have caused the gross opening and movement of the wall, leading to further tearing of the wall-floor plate around much of the circumference (Fig. 8.11).

- Improper penetration of welds in vertical full-penetration welds, due to a quality control issue during original construction. We do not have enough original design information or close-up inspection of the failed welds and accumulated corrosion to validate this hypothesis.

- Very high local ground motions (say PGA > 0.8g or so). We have no local site instrument to confirm this possibility, but the apparent modest levels of damage to four (partially filled) liquid fuels tanks and pump station building suggests that the local level of ground motion did not exceed PGA = 0.25 g to 0.35 g.

We were unable to gain access to carefully inspect the performance of the nearby liquid fuels tanks. However, their observed (from the site boundary, about 100 feet from the tanks) minimal damage suggests that they were designed at the same time as the water tank, using the same paint system and same staircase designs. The lack of damage to the somewhat larger (taller) liquid fuel tanks to the water tanks suggest that possibly they were no more than 50 percent full at the time of the earthquake, not an uncommon practice.

Wastewater Systems

Wastewater collection is provided by local communities. Essbio has a large collector pipe that takes the raw sewage to a wastewater treatment plant in Concepcion.

On February 27, 2010, Essbió was able to provide piped wastewater service for 92 percent of its customers in Region VI, 85 percent in Region VII, and 43 percent in Region VIII, with the largest outages in Concepcion. The primary causes of the wastewater outage in Concepcion were the failure of the main collector pipe due to imposed lateral spreads/PGDs and damage to baffles to the primary treatment tanks at the WWTP. This resulted in direct discharge into the Bio Bio River.

There was damage to a large-diameter wastewater pipe in Los Angeles, resulting in direct discharge into the Quilque River.

The last region to get wastewater service restored in Region VIII was likely about 4500 customers in Talcahuano–Ifarle. Wastewater was temporarily discharged into the Ifarle Canal as the earthquake and tsunami destroyed a wastewater pumping plant and the main grid of collector pipes in this area.

By April 10, 2010, Essbio was able to provide piped wastewater service for 99 percent of its customers in Region VI, 98 percent in Region VII, and 75 percent in Region VIII.

Canals

Water canals and hydraulic structures (levees, river embankment defenses) are used in many areas in Chile to deliver water from the Andes to communities in the central agricultural areas for irrigation purposes. These facilities were designed and are maintained by the MOP. As of April 1, 2010, the status of these facilities was as follows:

- 98 facilities were in the strong shaken area—43 in Biobio Region VIII, 26 in Maule Region VII, 13 in Santiago, 7 in O'Higgins, 5 in La Araucania, three in Valpariaso, one in Los Rios
- $2,111,000 had been spent for emergency repairs.
- $57,700,000 was projected for long-term repairs.
- 42 facilities were categorized initially as being in an extreme emergency condition.
- 32 facilities were categorized initially as being in a serious emergency condition. For example, damage to concrete linings in canals.
- 16 facilities were categorized initially as being in a moderate emergency condition. For example, minor damage that can be repaired under normal maintenance.
- 25 facilities collapsed, 38 had structural damage (ranging from light to extreme), three had nonstructural damage (ranging from moderate to extreme), one overflowed, two ruptured (extreme); eight had cracks (either extreme or serious); two were undermined, and 19 had other types of damage.
- 69 canals were being evaluated for long-term solution.
- 7 canals were being repaired.
- 14 canals had been repaired.
- 5 canals were under design.

Other Factors

Within 8 hours of the earthquake, there was general chaos in the Concepción area. This included looting of stores and setting fires. The local police could not maintain safety. It took 3 days for the military to show up in Concepción (there were no local military bases, and travel by road was nearly impossible due to bridge damage), after which crowd control was restored, often at gunpoint. The need for safety for Essbió staff and contractors as well as the loss of communications led to a slowdown in restoration efforts of at least several days.

Two fires were reported in the Concepcion area after the earthquake. Both fires were reported to have been arson. Almost none of the general building stock uses wood. No earthquake-ignitions (requiring fire department response) or fire spread occurred.

Essbió has a permanent staff of about 500 people. Essbió uses outside contractors for nearly all pipe work, with the common support of about 500 contractors. Before the earthquake, Essbió had four pipe repair crews. After the earthquake, Essbió retained additional pipe repair crews, reaching 70 crews system-wide during the peak pipe repair effort.

After the earthquake, to offset the loss of piped potable water supply in many areas, Essbió installed 900, 1000-litter tanks in affected zones to provide potable water to local communities. These tanks were filled at the water treatment plant and then trucked out to a suitable point in the distribution system. People took water from the tanks as needed via small hose bibs, filling canisters they brought with them. Essbió did not "man" these sites; instead, water usage was self-administered by the local residents.

Raw water turbidity in the Bió Bio River spiked after the earthquake, but returned to normal within a few days. The spike about the same as would be expected during a winter storm and did not have much influence on water quality.

The cell phone network in Concepcion suffered general failure and took about 5 days to restore. Essbió normally uses cell phones for communication with staff and contractors; it did not have a private radio system. The lack of communications slowed the restoration of water service, perhaps by 3 to 5 days overall.

Figure 8.4. Collapsed wall at Mochita WTP raw water pump station

Figure 8.5. Repairs being made to clarifier at Mochita WTP

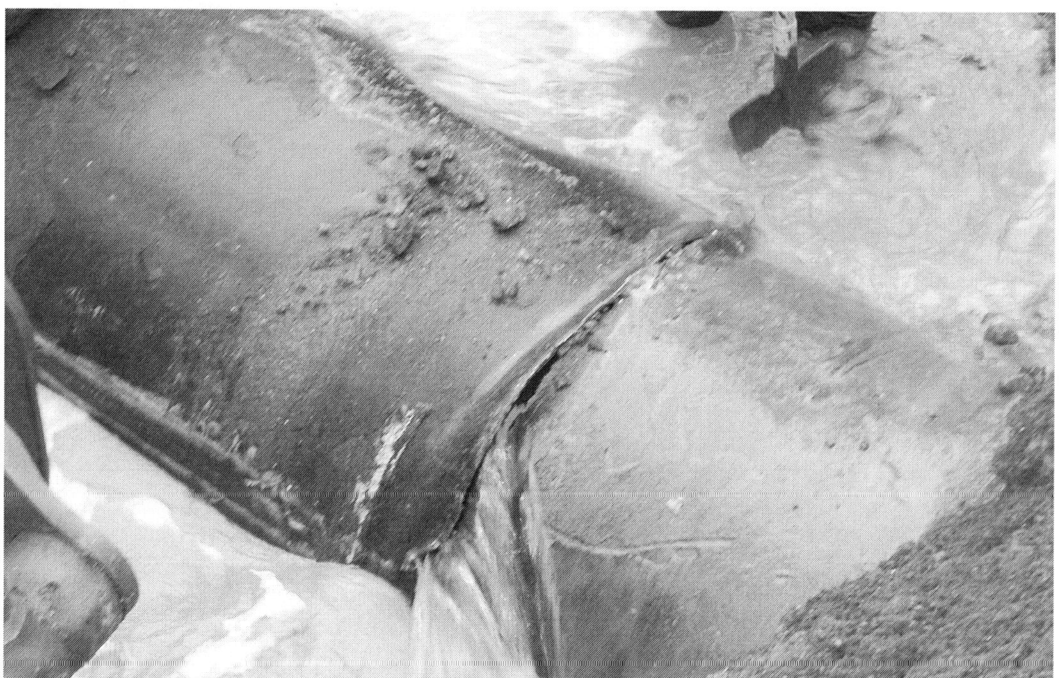

Figure 8.6. Damaged large-diameter welded steel water pipe

Figure 8.7. Typical elevated small steel tank (constructed in 2006)

Figure 8.8. Frame connection to a column (constructed in 2006)

Figure 8.9. Typical collapsed, elevated small steel tank (constructed in 1999)

Figure 8.10. Collapsed water (foreground left) and nearby liquid fuels tanks at Santiago Airport

Figure 8.11. Collapsed water tank at Santiago Airport

Figure 8.12. Roof-wall-staircase connection of collapsed water tank at Santiago Airport

8.2 Overview of System Performance

Potable water outages in Concepción were widespread and lengthy. Most of this was attributed to non-seismically-designed pipe failures due to PGDs. At least 73 elevated water storage tanks in rural water systems collapsed. Damage to baffles and settling tubes in clarifiers was common. Use of unreinforced brick/masonry/adobe infill walls contributed to structural damage of water treatment plant buildings as well as office buildings. Nonstructural damage occurred to desktop computers, laboratory equipment, and suspended ceilings. Lack of up-to-date emergency response and preparedness plans, spare parts, emergency generators, repair crews, and heavy equipment, with coincident failure of the cellular phone network and general chaos in the community in the first few days after the earthquake, all led to a slow restoration of service.

8.3 Damage Sites

Mochita WTP

Figure 8.13 shows settlement at the Mochita WTP that led to damage of a u-shaped concrete water channel.

Figure 8.13. Collapsed water tank at Mochita WTP

Figure 8.14 shows damage to the baffles at one of the four clarifiers at the Mochita WTP. Some tube settlers are displaced in the background. There was similar damage to all four clarifiers.

Figure 8.15 shows the damage to the pump station at the Mochita WTP. The building uses nonstructural infill walls (some of which cracked) and has non-ductile concrete columns. While the building did not collapse as a result of the earthquake, there was much concern that it might subsequently collapse in an aftershock. If the building had collapsed, it would have damaged the motors and pumps it housed, leaving Concepcion without a water source. Accordingly, Essbio installed steel braces and "catchers" around all the motors (Fig. 8.16).

Figure 8.17 shows 14, 1-ton chlorine tanks at the Mochita WTP. These are located inside a steel building. None of the tanks was anchored, yet we observed no evidence of the tanks sliding on their wood skid supports. This was confirmed by the plant staff. Had ground motions been PGA of 0.60 g or larger, there would almost certainly have been some sliding. While there was no strong motion instrument near the WTP, we relied on observed evidence to estimate the actual motions. Given the various levels of damage to nonstructural items (such as bookcases, windows, suspended ceilings, etc.), we estimate the actual horizontal PGA at the site to have been about 0.30 g to 0.45 g.

Figure 8.14. Damaged baffles at Mochita WTP

Figure 8.15. Damage to the pump station at Mochita WTP

Figure 8.16. Installed steel braces and "catchers" over pump station motors

Figure 8.17. Chlorine tanks at Mochita WTP

The filter building showed no evidence of structural or nonstructural damage. Figure 8.18 shows the operating deck floor; none of the control panels had sliding or damage to components within. The building style is reinforced concrete roof supported on columns and beams. We did not observe cracking in the structure. The filter boxes, immediately outside the windows in Figure 8.18, were not known to have any leakage post-earthquake, nor was there damage to the launders.

Figure 8.19 shows the pipe galley directly beneath the operating floor (Fig. 8.18). There were no leaks to any of the large-diameter pipes or damage to any valves or controls. Two-level, rod-hung cable trays line both sides of the filter gallery. These cable trays have no lateral supports as would be required by the nonstructural provisions of the ASCE 7, IBC, NBCC, SMACNA, and NSF13 codes and guidelines. This strongly suggests that similar code provisions that proscriptively require lateral bracing of cable trays, conduits, and pipes at about10-ft. spans need to be re-written. Note in Figure 8.19 that by providing some flexibility in the cables from the trays, the trays can swing laterally during the earthquake without imposing any distress on the rod hangers or the electrical cables.

Figure 8.20 shows a partially fallen suspended ceiling at the WTP. The fallen tiles and materials had no life-safety consequences, and the plant operators on shift during the earthquake reported no egress issues. This suggests that code provisions to require seismic design for suspended ceilings as a life-safety requirement are not entirely logical. Instead, it would make some sense to improve these types of ceilings over critical areas (such as control rooms) and use proper seismic detailing, while accepting lower-cost,

Figure 8.18. Filter building operating floor

Figure 8.19. Filter building pipe gallery

suspended-roof installations in other areas (warehouses, etc.). For life safety, it would appear sufficient to add wires to heavier items in roofs to prevent them from falling on people below. For locations where sprinkler heads project through the suspended ceiling, the underlying pipe-ceiling interaction needs to be considered, as impacts of the ceiling with some types of sprinkler heads can snap off the heads leading to unwanted water release and damage. In the United States, a wire-hung suspended ceiling might cost about $3 per square ft. to install; test data shows that modern code provisions for seismic suspended ceilings (wire diagonals plus compression struts) provides little or no capacity much beyond PGA = 0.4 g. The cost to build truly seismic suspended ceilings with complete lateral load paths with load-rated components can easily cost 10 times the basic cost of a suspended ceiling. Given the cost and performance issues and the common observation that suspended ceilings tend to fall apart at about PGA 0.30 g to 040 g, this suggests that the current codes and state of practice in the United States are resulting in extra cost for seismic bracing without delivering much in the way of extra seismic performance. To ensure good performance of suspended ceilings, a qualified engineer should verify the complete load path of the ceiling system.

In the United States, we would recommend that the startup batteries be properly restrained (there were none for this generator) and put in suitable snubbers to restrain the isolated generator. Neither of these improvements was in place at the time of the earthquake, yet the generator performed as required, even if an isolator was damaged (Fig. 8.21). This suggests that the level of ground motion did not exceed PGA = 0.50 g, as the batteries would almost certainly have slid or toppled at higher levels. Still, the relative low cost to restrain batteries (usually under $200) and snub a generator (usually under $2,000) seems like a good practice for such vital equipment at a WTP.

Figure 8.20. Damaged suspended ceiling

Figure 8.21. Emergency generator at Mochita WTP

Figure 8.22 shows some of the nearly 900 plastic tanks used to distribute potable water in and near Concepcion after the earthquake. The tanks were filled at the Mochita WTP, transported by truck, and dropped off at a suitable distribution point. Repairs are being made to two steel pipes. Temporary manholes are cut. The pipes are aligned with external bars, and then custom-made butt straps are installed. Common repair times for a crew, per steel pipe were about 48 hours (working 24 hours shifts) (not including disinfection or repairing the road).

Water Transmission Pipes

Figures 8.23 to 8.29 show some examples of damaged large diameter water pipes in Concepcion

A broken gate valve (flange failure) about 36 in. in diameter within a concrete vault was resolved by removing the valve and replacing it with a straight piece of pipe (Fig. 8.28). The lack of spare valves, parts, and the urgent need to restore water supply led to this type of repair.

Figure 8.22. Plastic tanks used for water distribution

Figure 8.23. Wrinkled steel pipe

Figure 8.24. Erosion due to large-diameter pipe failure

Figure 8.25. Failure of two parallel steel pipes at same location

Figure 8.26. Repair effort for two parallel failed steel pipes

Figure 8.27. Repairs completed

Figure 8.28. Broken gate valve and pipe repair in vault remnants

Figure 8.29. Another pipe break

Wastewater Treatment Plant

Figure 8.30 shows damage to the baffles at a clarifier at the Concepcion Wastewater Treatment Plant (WWTP). Figure 8.31 shows damage to the external concrete walls of the clarifier.

Figure 8.32 shows the pull-apart of the concrete walls in the chlorine contact tank. There were liquefaction-caused PGDs, and the level of damage suggested that it would take about 6 months to bring the plant back in service. One of the main sewers bringing raw sewage to the plant was broken near the Bio Bio River, so there was no influent to the WWTP, and raw sewage was flowing into the river.

Distribution Pipes

Figure 8.33 to 8.36 show typical repairs to common distribution pipes (6 in. to 8 in. common). The style of repair of using pipe clamps, inserting PVC pipes and such, shows the range of damage and repair strategies.

Figure 8.30. WWTP clarifier damage

Figure 8.31. WWTP clarifier damage

Figure 8.32. WWTP chlorine contact tank failure

Figure 8.33. Pipe repair

Figure 8.34. Pipe repair

Figure 8.35. Pipe repair

Figure 8.36. Pipe repair

8.4 Major Observations and Recommendations

1. As in many parts of the world, underground water pipes in Chile have been installed over the past 50 to 75 years with no seismic design provisions.
2. Some of the newest distribution pipe installations in Concepción (Essbio Water Company) used HDPE pipes. These appeared to have performed much better in this earthquake than other types of water pipe.
3. When installing large diameter (750-mm /30-in. diameter) welded steel pipe through potential liquefaction zones, it is important that girth welds are full penetration; otherwise, the pipes will not be ductile. The diameter-to-wall-thickness ratio should be no more than 95 in liquefaction zones.
4. In communities prone to liquefaction in Chile or elsewhere in the world, large earthquakes can damage buried pipes on a widespread basis. This will result in long water outages, with restoration of the last customer taking more than 30 days. In these areas, water companies serving communities with high economic activity, should consider adopting new seismic standards for installation of new water pipes.
5. The seismic design provisions in the American Lifeline Alliance (ALA) (2005) have been confirmed to be sound, based on observations in this earthquake.
6. Water and wastewater treatment plants that are located in liquefaction zones underwent moderate to extreme damage due to ground failure in this earthquake. This has been observed in past earthquakes around the world. Ideally, these facilities should be located outside such zones or must be designed to accommodate liquefaction.
7. Many older water utility buildings were heavily damaged due to inertial overloads. These could have been easily mitigated before the earthquake by suitable seismic upgrades.
8. Unbraced cable trays and unanchored chlorine one-tank tanks performed well in this earthquake. The cable trays, lacking any lateral supports, would have likely performed well at nearly any credible level of shaking, and we do not recommend unnecessary bracing of cable trays. Unanchored chlorine tanks should be restrained so that they are reliable at least PGA = 0.6 g in any high seismic zone worldwide.

8.5 Acknowledgements

Many people participated as part of the ASCE TCLEE Earthquake Investigation Team. This report would not have been possible without their efforts.

Mr. Jose Luis Arrano Urzua and Mr. Claudio Santelices Boettcher of Essbio provided access to facilities and provided technical and logistical information. Many other Essbio staff provided input, including staff at the water treatment plant (plant operators, water quality staff) and at headquarters (heads of maintenance, operations and engineering). Mr. Luis Estelle Aguirre (Government of Chile MOP, Santiago) and Mr. Carlos Sanhueza Sanchez (Government of Chile MOP, Concepcion) describe the performance of rural water systems and regional canals. All Essbio and MOP staff provided transparent descriptions of how the water and wastewater and hydraulic systems performed.

Mr. John Eidinger (G&E), Ms. Yumei Wong of DOGAMI, Dr. Leonardo Duenas-Osorio (Rice University), and Mr. Geri Pranzini (GHD) performed field reconnaissance at the Essbio WTP and attended meetings with Essbio officials. Mr. Anshel Schiff attended meetings with Essbio officials. Other members of the ASCE TCLEE team were Dr. Alexis Kwasinski (University Texas Austin), Dr. Bill Fullerton, Mr. Alex Tang, Ms. Allison Pyrch, Mr. Tom Cooper, and Mr. Roy Imbsen.

8.6 References

Boroschek, R., Soto, P., Leon, R., Maule Region Earthquake, February 27, 2010 Mw = 8.8, University of Chile, Faculty of Mathematics and Physical Sciences, Civil Engineering Department, Report 10/08, August, 2010.

9
AIRPORTS

Air traffic and airport operations were significantly affected by the moment magnitude (Mw) 8.8 earthquake off the west coast of Maule Region, Chile, which struck at 3:34 a.m. local time on February 27, 2010. In Santiago, the passenger terminal of the Comodoro Arturo Merino Benítez International Airport suffered significant nonstructural damage. If the earthquake had occurred during peak occupancy of the terminal, many injuries and casualties could have resulted. The air traffic control (ATC) tower sustained structural damage and was rendered inoperable, forcing the activation of emergency, provisional measures. Important support structures were also affected.

Region-wide, communications were compromised, including the Santiago Area Control Center, which has the responsibility of controlling aircraft en route to airports in Santiago and Concepcion, the largest cities near the earthquake's epicenter and an important logistics hub for the relief effort.

Carriel Sur International Airport in Concepción was only slighted affected by the earthquake. The terminal building suffered water damage due to broken sprinkler system. In the control tower, a few panes of glass windows were broken and electronic equipment was dislocated. This physical damage did not significantly affect relief operations. However, as flight operations increased sevenfold above normal during the relief effort, the stress on the air traffic controllers of this lifeline link became evident.

The airport at Los Angeles was closed due to extensive damage to the control tower. Only emergency medical and relief supply flights were allowed.

9.1 Description of Airports

Members of the ASCE/TCLEE post-earthquake lifeline performance evaluation team visited three airports in the earthquake-impacted area. Santiago International Airport is the farthest away from the epicenter at about 330 km northeast. Concepcion International Airport is the closest at about 100 km south, and Los Angeles Airport is about 170 km southeast of the epicenter.

Santiago International Airport (33.396997° S, 70.793310° W)

Comodoro Arturo Merino Benítez International Airport (airport code SCL), also known as Santiago International Airport, is the country's largest airport. It is Chile's international gateway, serving more than 9 million passengers in 2009.

SCL has two parallel runways on east and west side of the terminal building, modern passenger and cargo terminal facilities, a military base, and typical airport support infrastructure. Both runways are equipped with an instrument landing system (ILS). Ground, tower, and departure/approach services are provided on a 24-hour basis from the

ATC tower located between the two runways to the north of the passenger terminal. A backup control tower, maintained for contingency purposes, is located at the midpoint of the older runway (17L/35R)—the east runway, close to the fire department.

Concepción International Airport (36.77796° S, 73.05904° W)

Carriel Sur International Airport (airport code CCP) is located 5 km northwest of Concepción, approximately 450 km southwest of Santiago. It is the second largest airport in Chile and serves as a primary alternate airport for SCL. CCP was constructed under the Alliance for Progress Program, in part because of national interest in developing an airport in Chile's southern region to respond to natural disasters such as the devastating 1960 Valdivia earthquake. Despite being the second most important airport in Chile, Carriel Sur does not have radar service for approach/departure control. The approach/departure control service is provided using non-radar protocols, in coordination with Santiago's Area Control Center. Carriel Sur's control tower provides ground, tower, and approach services on a 24-hour basis and is staffed by three air traffic controllers at all times. Runway 02 is supported by a category I ILS.

Carriel Sur became the most important hub for relief supply after the earthquake due to its proximity to the immediate disaster area.

Los Angeles Airport (37.402376° S, 72.421746° W)

The Los Angeles, Chile, Maria Dolores Airport (airport code LSQ) is a domestic airport with one runway for small aircraft. It is located about 5 km northwest of the city.

The control tower is constructed as part of the passenger terminal building. The service building that houses a backup power generator and termination of commercial power feed is located south of the terminal building.

This airport became an important part of the lifeline system after the earthquake due to its location within the affected area. Medical resources, emergency supplies, and equipment were delivered through this airport.

9.2 Overview of System Performance

Both Santiago and Concepcion Airports were open 2 to 3 days after the earthquake due to various problems reported in this section. Los Angeles Airport remained closed until the Civil Aviation Authority could evaluate the damage and decide what measures to take before re-opening the airport. However, the airport was open to authorized relief supply and military flights. Common failure modes observed in all three airports visited are

1. control tower damage (both structural and non-structural),
2. passenger terminal damage (both structural and non-structural), and
3. loss of communication.

The airport system experienced both structural and nonstructural damage. An airport system consists of passenger and cargo terminal buildings, an air traffic control tower, communication equipment (including traffic control equipments), the runway system, fuel supply system, electric power system, and security and emergency services associated with its operation.

The collapse of the water tank for firefighting at Santiago International Airport might have contributed to the delay in re-opening the terminal.

Although there were no casualties at the airports due to the timing of the earthquake, the damage to the airport system was quite significant.

9.3 Santiago International Airport (SCL)

Santiago International Airport sustained extensive damage to its passenger terminal building, the traffic control tower, the walkways connecting the viaduct overpass to the departure level, and the airline service buildings. A major regional carrier, LAN, experienced extensive damage to one of the office buildings within the airport. Due to the time of the earthquake, the airport terminal was not open. The extensive nonstructural damage within the passenger terminal did not result in any injury or death.

The 1.4 million-liter steel water tank that provides water for firefighting collapsed. Luckily, there was no fire at the terminal after the earthquake; however, there was no damage to either runway.

Figure 9.1 provides a brief overview of the layout of Santiago International Airport. The office buildings are located on the east and west end of the passenger terminal.

Santiago International Airport Terminal Building

The SCL passenger terminal building suffered widespread nonstructural damage, which would likely have caused serious personal injury had the earthquake occurred during peak occupancy of the passenger terminal building (PTB). In the departure hall (about 300 m long by 60 m wide) HVAC equipment and ductwork in the super structure overhead were damaged, and strong shaking brought down the plastic panels of the false ceiling throughout the PTB. A number of pipes in rest rooms and around departure gates areas were broken. Light fixtures, suspend ceilings, wiring, air exit vents, air ducts and air pump units fell. Many glass windows and glass walls were broken or shaken loose. The elevators were shaken off their tracks or the counter weight dislocated from its guides.

In addition to fallen suspend ceiling tiles, large overhead light fixtures in bars and restaurants in the terminal building fell to customer seating areas.

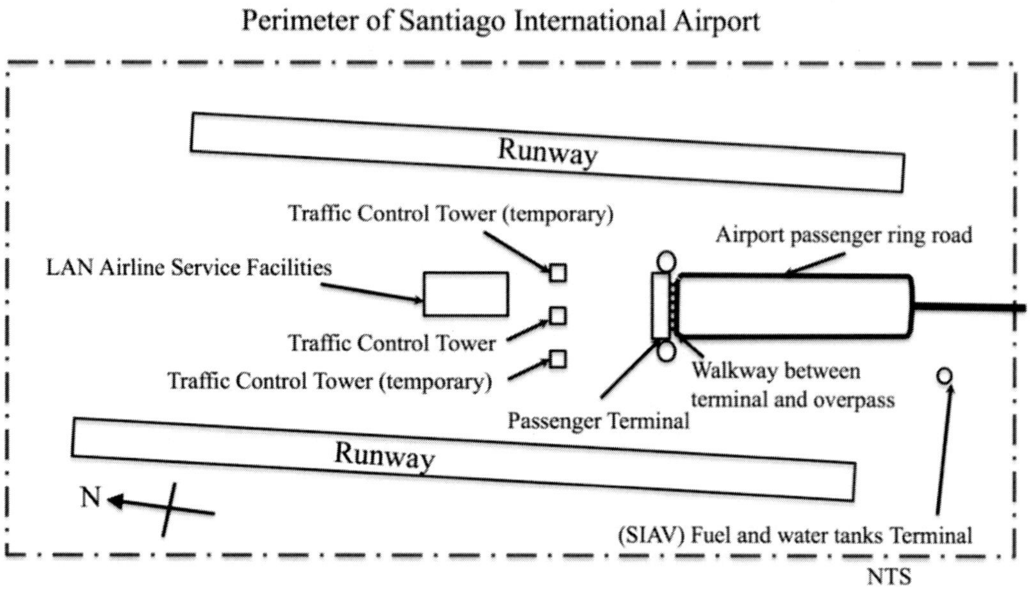

Figure 9.1. Layout of the facilities of the Santiago International Airport (SCL)

The widespread nonstructural damage in the PTB resulted in delayed opening of the airport. Many businesses operations in the PTB also suffered both revenue and property losses. Some were not open for business until well after the airport terminal re-opened.

Figures 9.2 to 9.18 show the extensiveness and the large scale of damage sustained by the passenger terminal building and the business operations in the building. It is easy to visualize injuries to passengers if the earthquake had occurred during operating hours.

The office wings on both ends of the terminal sustained failure modes similar to those observed in the passenger areas. Figures 9.19 to 9.20 show typical failures.

All structural damage was minor and repairable and posed no threat to life safety (Fig. 9.21 to 9.25). The damage observed is to the steel structures on both the outside and inside of the terminal building.

One of six passenger pedestrian bridges between the viaduct curbside and the terminal building had to be replaced. All bridges sustained damage to support connections, both on the viaduct curbside and the terminal side. All were replaced with temporary structures pending a permanent, seismically sound repair or were removed before the TCLEE team visit. At the time of the visit, these walkways were the main evacuation route from the terminal building's upper level. Figures 9.26 to 9.32 show the damage and the temporary repairs. One section of the canopy on the viaduct curb toppled (Fig. 9.33) and was later removed before re-opening of the airport facilities.

Figure 9.2. Hundreds of ceiling tiles fell and covered the whole international departure check-in level. (Courtesy of Alejandro Kolbe V., Dirección de Aeropuertos - Ministerio de Obras Públicas, Chile)

Figure 9.3. The domestic departure area looked the same as in the international departure area. (Courtesy of Alejandro Kolbe V., Dirección de Aeropuertos - Ministerio de Obras Públicas, Chile)

Figure 9.4. The group check-in area in the domestic flight check-in area, the fallen air conditioning unit could cause serious injury to check-in passengers. (Courtesy of Alejandro Kolbe V., Dirección de Aeropuertos - Ministerio de Obras Públicas, Chile)

Figure 9.5. Ground floor arrival area sustained damage similar to the departure level. (Courtesy of Alejandro Kolbe V., Dirección de Aeropuertos - Ministerio de Obras Públicas, Chile)

Figure 9.6. Metal channels and the pipes fell from the ceiling structure in the departure gate aisle. (Courtesy of Alejandro Kolbe V., Dirección de Aeropuertos - Ministerio de Obras Públicas, Chile)

Figure 9.7. An air condition unit fell through the ceiling in the flight boarding area leading to the skyway to the plane (Courtesy of Alejandro Kolbe V., Dirección de Aeropuertos - Ministerio de Obras Públicas, Chile)

Figure 9.8. Air conditioning unit fell through the ceiling in another waiting area for boarding. (Courtesy of Alejandro Kolbe V., Dirección de Aeropuertos - Ministerio de Obras Públicas, Chile)

Figure 9.9. An air conditioning unit fell on the aisle of the boarding gate area and in front of a duty free shop. Note the ceiling metal bars also fell along this aisle. (Courtesy of Alejandro Kolbe V., Dirección de Aeropuertos - Ministerio de Obras Públicas, Chile)

Figure 9.10. The pipes fell from the ceiling in the food court area along the window; the water leak indicates that some pipes are broken. (Courtesy of Alejandro Kolbe V., Dirección de Aeropuertos - Ministerio de Obras Públicas, Chile)

Figure 9.11. Cool air vents fell from the ceiling, a common failure throughout the terminal building. (Courtesy of Alejandro Kolbe V., Dirección de Aeropuertos - Ministerio de Obras Públicas, Chile)

Figure 9.12. Ceiling metal bars fell onto the ramp leading to the skyway for boarding. (Courtesy of Alejandro Kolbe V., Dirección de Aeropuertos - Ministerio de Obras Públicas, Chile)

Figure 9.13. Broken glass partitions occurred in many areas within the terminal building. (Courtesy of Alejandro Kolbe V., Dirección de Aeropuertos - Ministerio de Obras Públicas, Chile)

Figure 9.14. An elevator cab dislocated from its tracks and counter balance weight and disengaged from the guide. All the elevators were required to be repaired before putting back to service. (Courtesy of Alejandro Kolbe V., Dirección de Aeropuertos - Ministerio de Obras Públicas, Chile)

Figure 9.15. Light fixtures toppled in a restaurant. (Courtesy of Alejandro Kolbe V., Dirección de Aeropuertos - Ministerio de Obras Públicas, Chile)

Figure 9.16. The light fixture in this bar fell in the customer seating area. (Courtesy of Alejandro Kolbe V., Dirección de Aeropuertos - Ministerio de Obras Públicas, Chile)

Figure 9.17. Both the inside and the outside of this convenient store were covered with materials that fell from the ceiling structure. (Courtesy of Alejandro Kolbe V., Dirección de Aeropuertos - Ministerio de Obras Públicas, Chile)

Figure 9.18. Toppled gumball machine in hallway (Courtesy of Alejandro Kolbe V., Dirección de Aeropuertos - Ministerio de Obras Públicas, Chile)

Figure 9.19. The pipes and ceiling fixture fell to the floor littering the aisle and exposing air conditioning ducts and sprinkler pipes. (Courtesy of Alejandro Kolbe V., Dirección de Aeropuertos - Ministerio de Obras Públicas, Chile)

Figure 9.20. Almost all ceiling tiles fell along with attached fixtures. (Courtesy of Alejandro Kolbe V., Dirección de Aeropuertos - Ministerio de Obras Públicas, Chile)

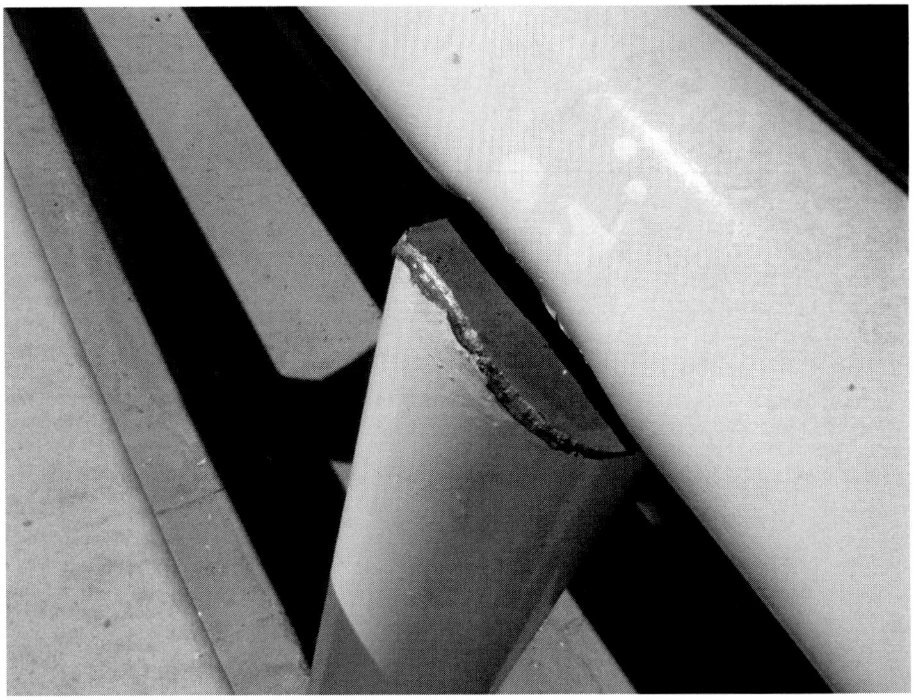

Figure 9.21. Weld separated between the column and the beam, most likely due to poor weld. (Courtesy of Alejandro Kolbe V., Dirección de Aeropuertos - Ministerio de Obras Públicas, Chile)

Figure 9.22. Bolted joints between beam and column where the bolts were sheared; there was no collapse as the beam was still sitting on the bracket that was welded to the column. (Courtesy of Alejandro Kolbe V., Dirección de Aeropuertos - Ministerio de Obras Públicas, Chile)

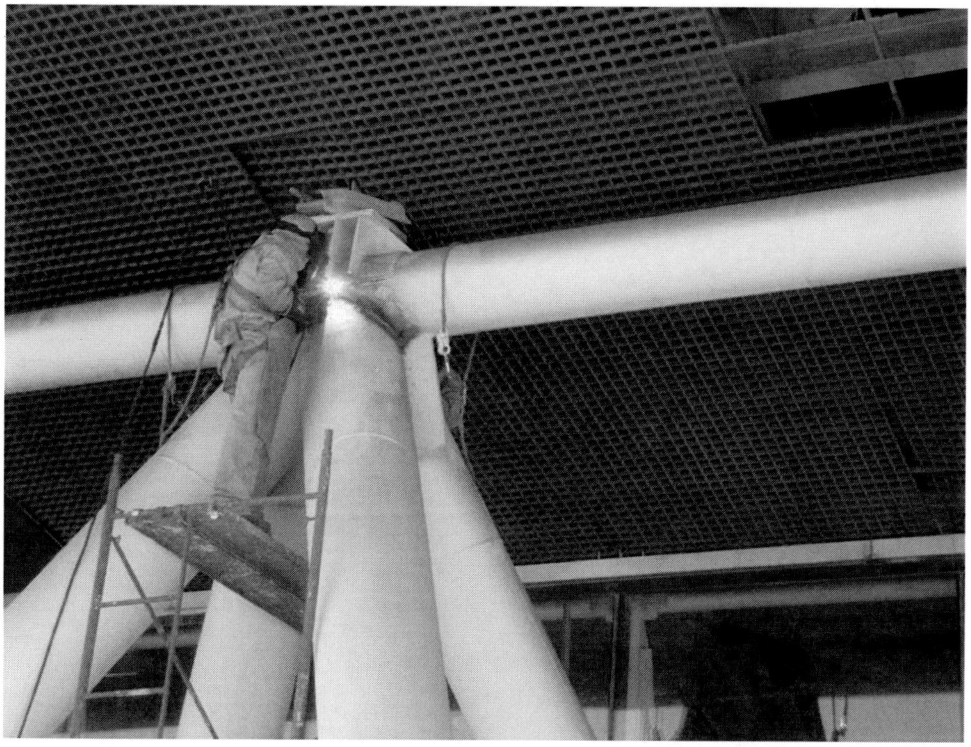

Figure 9.23. This repair was done immediately to ensure safety. (Courtesy of Alejandro Kolbe V., Dirección de Aeropuertos - Ministerio de Obras Públicas, Chile)

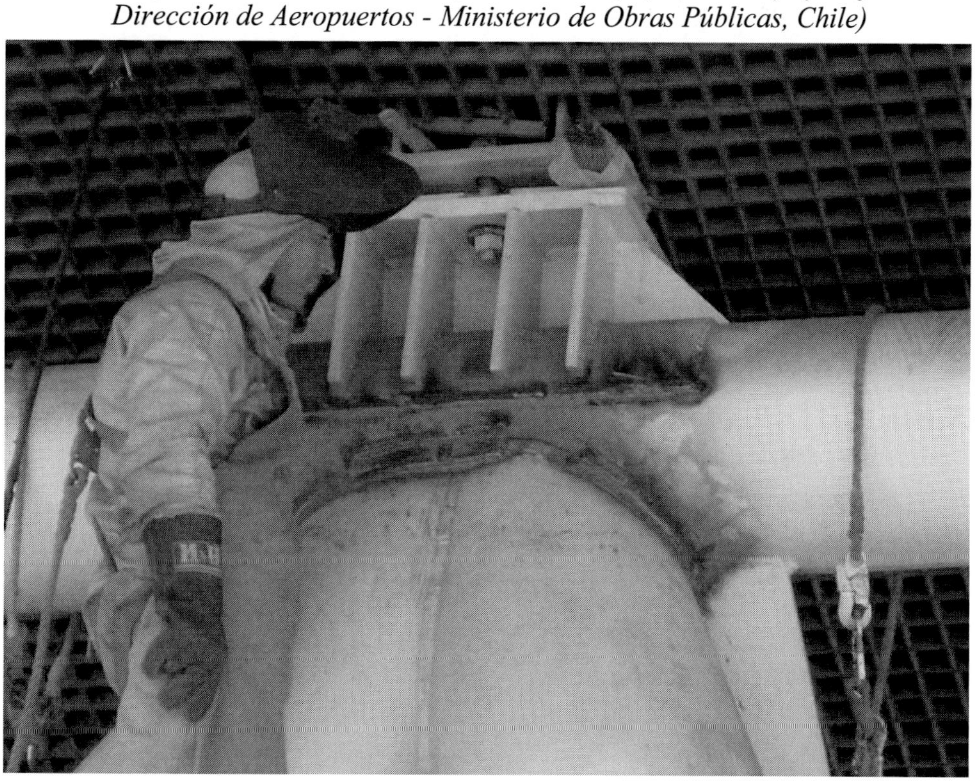

Figure 9.24. The finished repair joint with a small plate added to strengthen the connection between the column and the beam. (Courtesy of Alejandro Kolbe V., Dirección de Aeropuertos - Ministerio de Obras Públicas, Chile)

Figure 9.25. Damaged connection of the steel frame that supports the roof of the passenger terminal building.

The terminal was closed and commercial passenger operations were practically suspended between February 27, 2010 and March 1, 2010. During that time, the passenger flights en route to Santiago had to be rerouted to airports outside Chile. On March 2, a large tent was installed to serve as a passenger terminal, which allowed LAN to partially re-launch operations (Fig. 9.34). On about March 5, part of the airport's domestic passenger terminal became operational for international passenger operations, allowing for a gradual increase in passenger flights. Large tents were used to process departing and arriving passengers. Immigration, customs, check-in, and baggage handling were processed in the tents. All departures and arrivals were set up on the west end of the terminal at ground level. Few flights were allowed until the airport terminal was safe to re-open in mid-March.

As the weather is still cool in March and April, the HVAC system had not yet been repaired at the time of the TCLEE team visit. Due to extensive damage, the airport authority wanted to explore all options for mounting HVAC equipment to ensure a better seismic design to mitigate future earthquake losses.

One of the 18 skyways located on the east side of the terminal building was damaged. Figure 9.35 shows the gangway with the skyway removed.

Figure 9.26. One of the six walk bridges between the viaduct curbside and the terminal building with the bearing pads broken on the viaduct side and fell. (Courtesy of Alejandro Kolbe V., Dirección de Aeropuertos - Ministerio de Obras Públicas, Chile)

Figure 9.27. One side of the bridge was disconnected from the steel frame bearing surface; the other side was still connected, and the bridge did not collapse. All five walk bridges failed on the viaduct end. (Courtesy of Alejandro Kolbe V., Dirección de Aeropuertos - Ministerio de Obras Públicas, Chile)

Figure 9.28. The bearing area of the pedestrian bridge was insufficient. This one showed pounding damage, but it did not collapse. All five damaged pedestrian bridges were replaced with temporary pedestrian bridges before re-opening of the airport. (Courtesy of Alejandro Kolbe V., Dirección de Aeropuertos - Ministerio de Obras Públicas, Chile)

Figure 9.29. The terminal side of the pedestrian bridge bearing structure after the pedestrian bridge was removed. A narrower temporary pedestrian bridge provides access to departure level after the airport re-opened.

Figure 9.30. The temporary pedestrian bridge

Figure 9.31. The collapsed pedestrian bridge was removed and the entrances for both levels were closed off.

Figure 9.32. The entrance of the collapsed pedestrian bridge after it was removed (as viewed from the terminal). (Courtesy of Alejandro Kolbe V., Dirección de Aeropuertos - Ministerio de Obras Públicas, Chile)

Figure 9.33 One section of the canopy on the curb side of the viaduct toppled and was removed before re-opening of the airport. (Courtesy of Alejandro Kolbe V., Dirección de Aeropuertos - Ministerio de Obras Públicas, Chile)

Figure 9.34. Large tents were set up to process passengers during the first two weeks after the earthquake.

Figure 9.35. A boarding gate with a damaged skyway at SCL

Air Traffic Control Tower (33.392089°S, 70.793929°W)

The ATC suffered both structural and nonstructural failures, including the ejection of the elevator counterweight. The four mounting pedestals of the steel frame that provides external support of the cab of the control tower to the concrete structure were damaged. Preliminary evaluation suggested that the edge of the anchors were too close to the concrete pedestal. The damage to the anchors and the concrete pads indicate torsional movement of the tower during strong shaking. The strong shaking also damaged equipment in the cab or caused it to fall on the floor. One wood equipment cabinet was lifted off its base despite being secured to a base that was anchored to the concrete floor. Only two glass windows were broken. The elevator had not yet been repaired during the team's visit in April, and we had to climb up the stairs to get to the control tower cab. Figures 9.36 to 9.40 show damage inside and outside of the control room. The scaffold on the outside of the tower was used to inspect the tower after the earthquake.

The raised floor of the control tower cab was undamaged; the floor tiles were screwed to the pedestal heads of the raised floor pedestals (Fig. 9.38).

An H-shaped concrete tower supports the traffic control cab (Fig. 9.41). The offset of the middle section from the centerline of the two members may have generated the rotation mode during strong shaking.

Figure 9.36. Two identical steel frames supporting the control tower cab

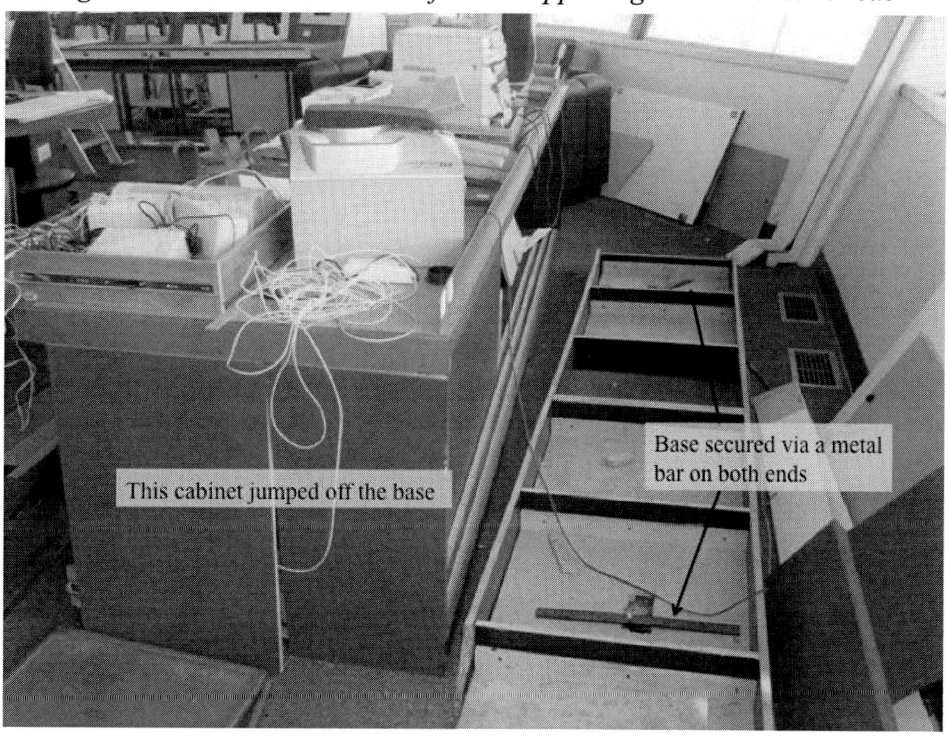

Figure 9.37. One of the cabinets for desk top equipment jumped off the base. Most of the equipment on desk tops fell on the floor. The equipment was not secured to the desk top.

Figure 9.38. The raised floor (about 60 cm/8 in. high) of the control tower cab where cabling are routed.

Figure 9.39. The typical failures observed in the control tower cab (Courtesy of Alejandro Kolbe V., Dirección de Aeropuertos - Ministerio de Obras Públicas, Chile)

Figure 9.40. Damage to the four connection points of the steel frames. The bending of the threaded rod in the top right figure indicated a rotating motion. The rust on the stud indicates that nuts were not installed.

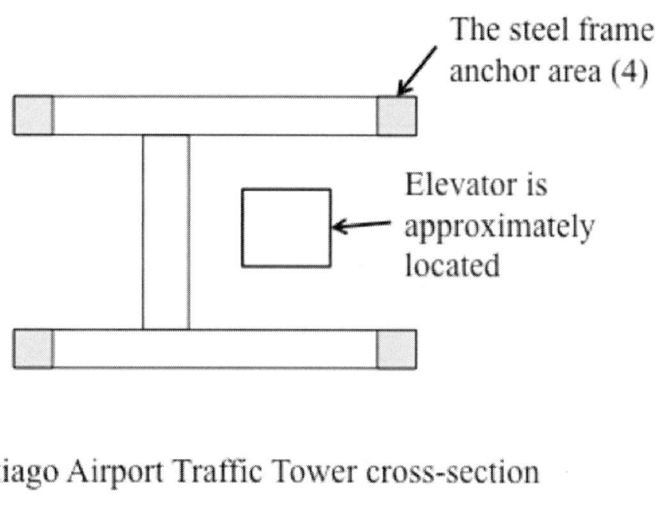

Figure 9.41. Cross section of the traffic control tower structure (NTS)

Immediately after the earthquake, the traffic control was moved to a provisional tower, which is normally used by the fire department and emergency response services (Fig. 9.42). The controllers took their portable radios and other required equipment to monitor required frequencies and re-direct arriving flights to airports north of Chile.

The runways were inspected, and runway 17L/35R was cleared for operations by 3:53 a.m., about 19 minutes after the earthquake struck. At 4:34 a.m., the first flight, an ambulance flight, landed. At around 2:00 p.m., the first commercial flight landed. It was a LAN flight from Auckland, which had done a technical stop at Isla de Pascua upon notification of the earthquake.

The radar system is located on the hilltop west of the airport, which did not sustain any damage. Because of the upcoming Feria Internacional de Aire y Espacio (FIDAE) event scheduled to begin on March 23, the airport needed to have both runways operational as soon as possible. Obtaining and installing a second provisional tower in time required too much time. Consequently, a temporary tower was quickly designed and installed at the midpoint of runway 17R/35L using a provisional cab mounted on two, stacked 20-ft. containers (Fig. 9.43).

Electric power to the airport was disrupted, but the uninterrupted power supply (UPS) system with diesel generator sets provided emergency back-up power automatically. There are three UPS systems serving the airport—one for the passenger terminal, one for the control tower and one for fire department and emergency services.

At the time of the TCLEE team's visit, both provisional towers were still in use. The main ATC tower was closed pending the necessary repairs and improvements to the cab mounting system. The new design had already been completed by the consulting firm Arcadis and was under review by the Ministry of Public Works.

Ancillary facilities were also affected. The fuel storage and distribution system lost the water tank for fire fighting, which collapsed. The leak control system also sustained some damage, which prompted authorities to shut down the water system for some days after the earthquake. The collapsed tank was an at-grade unanchored 1.4 million-liter welded steel water tank (Fig. 9.44). The failure was likely due to repeated wall uplifts and wall buckling by sloshing. The tank was reportedly only half full.

LAN Airline Offices and Facilities (33.391977°S, 70.792532°W)

The LAN Airline operations and maintenance facilities, approximately 200 m (600 ft.) northeast of the ATC tower, suffered extensive structural damage to one of the three buildings—one hanger and two office buildings (Fig. 9.1). Although the buildings are close together, they performed differently. The area has very poor soil condition. The two lightly stressed structures (the hanger and the new office building) had their foundations treated. The structure with extensive damage did not have its foundation treated.

The 21 year old maintenance hangar showed few signs of distress, with only the loss of some light fixtures and glass panes (Fig. 9.45 and 9.47). An Airbus 340 aircraft was undergoing maintenance in this hanger. The aircraft "walked" on its jacks approximately 1 m from its original position, evidence of the movement that the building withstood (Fig. 9.48).

A newly commissioned building, housing LAN Airline's administrative functions, also performed well with little sign of distress (Fig. 9.49). However, the 10-year old operations building was severely damaged. The diagonal steel bracing members had bent and in some cases sheered. Although the building did not collapse, the elevator counterweight was ejected and interior spaces were severely affected and rendered unusable. At the time of the TCLEE team visit, a contractor was replacing the external steel supports with new elements of the same design. Figures 9.50 to 9.51 show the damage sustained by this building and the repair being done.

LAN Airline was operating only about 85 percent of its normal flight schedule 6 weeks after the earthquake.

Figure 9.42. Temporary fire department control tower for the east runway

Figure 9.43. Equipment arrived second week after the earthquake, and this structure became the temporary tower for the west runway. The radar equipment for this airport is located on the hill to the east of the airport.

Figure 9.44. Collapsed water tank and, to the right in the containment area, fuel tanks.

Figure 9.45. The hanger doors were closed during the earthquake, as the Airbus 340 was being serviced.

Figure 9.46. The hanger and the aircraft in it survived the earthquake.

Figure 9.47. Damage to the office at the northeast corner of the hanger indicates strong shaking.

Figure 9.48. Another indication of strong ground motion at this location is the movement of one of the tripods that supported the aircraft.

Figure 9.49. LAN Airline administration office building in Santiago International Airport, one of the three buildings that did not sustain any damage.

Figure 9.50. LAN Airline administration office building, scaffolding to repair external brace members.

Figure 9.51. Damaged members placed on the sidewalk waiting for disposal

Fuel and Water Storage Tanks (33.409945°S, 70.795985°W)

The fuel and water storage tanks are located south of the terminal building, on the west side. This tank farm is owned and operated by Sociedad de Inversiones de Aviacion, Ltda (SIAV), which is responsible for planning and fuel supply for SCL. There are six tanks at this facility: Five are fuel tanks located in the containment area; the water tank for firefighting is located just west of the guardhouse and the equipment area (Fig. 9.52). Due to the collapse of the 1.4 million liter (350,000 gallons) water tank (Fig. 9.44), the northwest corner of the nearby masonry wall was damaged with the concrete frame remaining (Fig. 9.53).

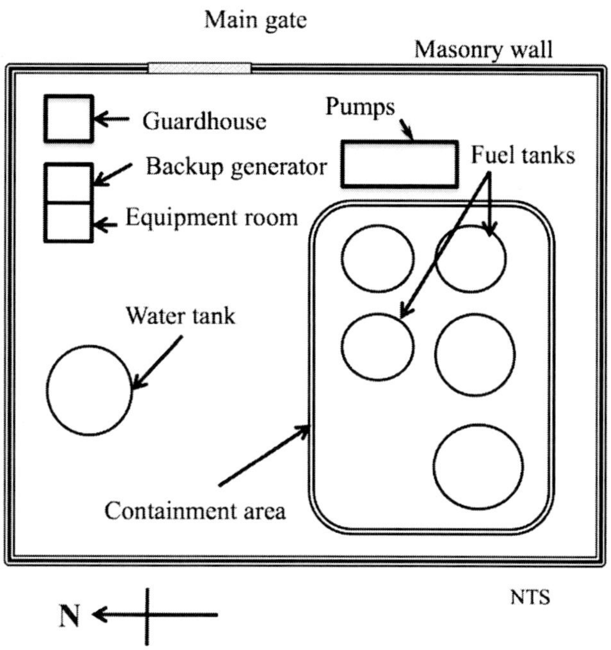

Figure 9.52. Layout of SIAV tank farm

Figure 9.53. Damaged masonry wall of the tank farm

Emergency Response

As the airport was essentially closed when the earthquake occurred; there was no immediate need of extensive emergency response requirement, except for the essential services on duty, such as ACT, terminal security, cargo handling, and airline maintenance.

With cellular phone service disrupted, communication between the control tower and the terminal was difficult. Staff hand carried messages back and forth immediately after the earthquake. There was confusion about getting damage reports out until a few hours after the earthquake. As previously noted, air traffic controllers implemented emergency procedures and immediately took working equipment and relocated to a designated area to monitor the air space. No accidents occurred, there was no fire, and no one was injured within the airport premises.

9.4 Concepción International Airport (CCP)

CCP suffered light damage to the administrative building and ATC tower, with plaster cracking evident as well as the loss of some false ceiling tiles. The cab itself lost three windows and its interior lighting; however, the consoles and equipment were in good condition. This structure was the original passenger terminal constructed in the 1960s (Fig. 9.54). The new passenger terminal, under private concession, suffered slight damage (Fig. 9.55). However, the building sprinkler system was activated by the shaking and flooded the building for almost 4 hours until maintenance personnel familiar with the system arrived to shut it off. Figure 9.56 shows the airport layout. CCP has only one runway and serves as the backup for Santiago International Airport.

Communications to the Santiago air control center (ACC) were disrupted initially. The ACC itself was not affected; however, communications throughout the region were degraded. Since the CCP ATC tower is not equipped with radar, it relies on communications with the ACC to manage air traffic approaching the airport. In the absence of this communication link, controllers had to manage all arriving traffic through pilot communications.

The chief controller lives near the airport and came to work with his neighbor, the logistics chief, once the initial shaking had subsided. Power was provided to the airfield through back-up generators. By 7:00 a.m., enough daylight was available to thoroughly inspect the runway and begin cleaning up the controller cab. There were still no regular communications with Santiago ACC; however, staff managed to get a cell phone message back with a brief condition report.

In the morning, Navy flights began to operate. Later that first day, government, general aviation, and helicopter reconnaissance flights resumed. ILS was taken out of service, as ground displacement could have affected the propagation of the signals to aircraft. Operations were visual, with a 1,000 m visibility requirement. A calibration flight from Santiago arrived in the first few days after the earthquake to restore the system.

Figure 9.54. Concepción Airport administrative building including the ATC tower. This structure served as the passenger terminal building before the new one was built.

Figure 9.55. The east end of the Concepción Airport passenger terminal building

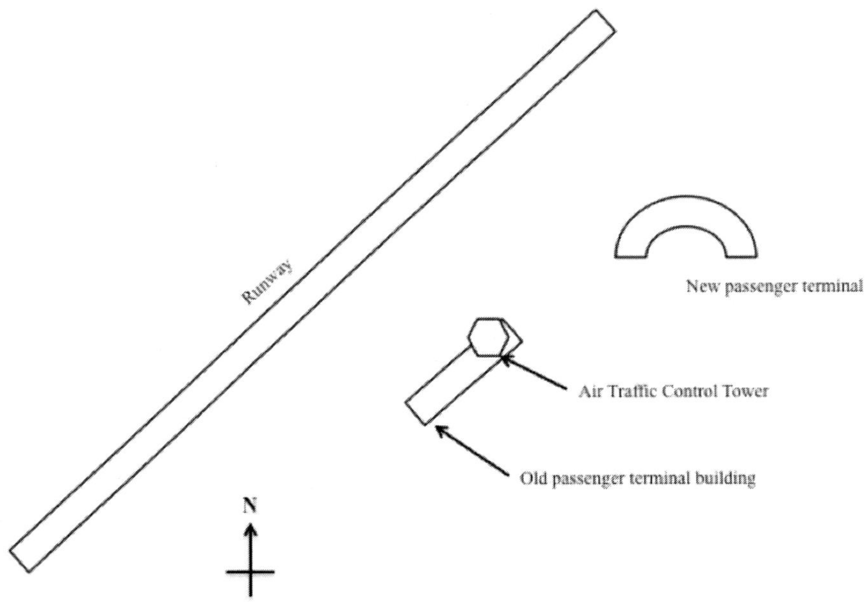

Figure 9.56. Layout of Concepción Airport

Air traffic operations quickly increased, peaking at 340 operations in one day—a sevenfold increase over normal operations. This traffic increase placed high stress on the airport systems and its personnel. The limited fuel supply led the airport to restrict refueling of aircraft. ATC operated six controller positions (compared to its normal two controllers) plus a supervisor. Controllers slept at the airport. They maintained a high esprit-de-corps, aware of the important role they were playing in the relief effort. However, the stress of maintaining safe operations under these conditions did affect the staff. The supervisor was ultimately placed on medical rest.

Terminal Building

The terminal lobby has a high ceiling with two levels at the back facing the airfield. The ground level serves both arriving passengers and ticket counters. The front is lined with glass windows and doors.

The terminal building was closed for about a day and a half due water and nonstructural damage, such as fallen ceiling tiles, broken pipes, and broken glass doors and windows. The water damage was caused by broken sprinkler heads in the main section of the terminal, whose connections were broken by the strong shaking of rigid wood ceiling planks during the earthquake. No one at the terminal knew where the main shut off was located; therefore, the water was allowed to run for hours.

Figures 9.57 to 9.64 provide a sense of the extent of damage to the terminal after the earthquake.

Figure 9.57. Damaged wood ceiling planks; many of them activated the sprinkler heads and caused flooding. (Courtesy of Rodrigo A. Esquivel Opazo of Engineering Consultant Victor Faraggi H.)

Figure 9.58. Disengaged back panel of one of the flight schedule display monitors. (Courtesy of Rodrigo A. Esquivel Opazo of Engineering Consultant Victor Faraggi H.)

Figure 9.59. Broken glass panel at the front of the passenger terminal (Courtesy of Rodrigo A. Esquivel Opazo of Engineering Consultant Victor Faraggi H.)

Figure 9.60. False ceiling damage above a small refreshment counter. (Courtesy of Rodrigo A. Esquivel Opazo of Engineering Consultant Victor Faraggi H.)

Figure 9.61. Broken and disengaged skyway entrance doors. (Courtesy of Rodrigo A. Esquivel Opazo of Engineering Consultant Victor Faraggi H.)

Figure 9.62. Airport manager's office after the earthquake. (Courtesy of Rodrigo A. Esquivel Opazo of Engineering Consultant Victor Faraggi H.)

Figure 9.63. Broken glass panel and water damage. (Courtesy of Rodrigo A. Esquivel Opazo of Engineering Consultant Victor Faraggi H.)

Figure 9.64. Exterior water leakage from the terminal building. (Courtesy of Rodrigo A. Esquivel Opazo of Engineering Consultant Victor Faraggi H.)

Air Traffic Control Tower

There was no structural damage to the tower. Three windows were broken with very minimal damage to the equipment in the control tower. Figures 9.65 to 9.68 show the condition inside the control tower after the earthquake.

Emergency Response

No injuries were reported at this airport. After communications with SCL resumed, the airport started to receive relief supply flights. Passenger traffic did not reopen until a week after the earthquake.

Dedicated ATC controllers managed to maintain a very high level of traffic during the relief effort.

9.5 Los Angeles Airport (LSQ)

The Los Angeles Airport (LSQ), also known as the Maria Dolores Airport, is located west of Los Angeles City. This very small airport with one runway, which is for small aircrafts, was not damaged. Limited numbers of commercial flights take off and land here. The ATC tower and the passenger terminal are colocated in a wooden structure (Fig. 9.69). To its south side is the equipment and maintenance personnel building. The building has a backup power generator and storage for spares, with a few rooms for offices. After the earthquake, due to damage to the passenger terminal, the director moved her office to this small building (Figure 9.70).

Terminal

There was nonstructural damage to the airport terminal; however, there were no broken glass panels. The wood frame and glass panels were dislocated from the beam slightly (Fig. 9.71). Ropes were used to keep the frame from collapsing in the event of a strong aftershock (Fig. 9.72). Some light fixtures were dislocated slightly, and a few cracks appeared on the plaster walls (Fig. 9.73 to 9.74). A few characters fell off the airport sign on the outside of the building facing the airfield (Fig. 9.75).

Figure 9.65. Air traffic control tower after the earthquake—a few broken windows, a fallen false ceiling, and shifting of cabinets and desks. (Courtesy of Ian Pickenpack, ATC Concepción)

Figure 9.66. Debris at the stairs leading to the control tower room. (Courtesy of Ian Pickenpack, ATC Concepción)

Figure 9.67. ATC supervisor standing with controllers and staff in the cleaned up room. The glass windows were not yet repaired. (Courtesy of Ian Pickenpack, ATC Concepción)

Figure 9.68. Flags of the countries providing relief flights were displayed outside the ATC. (Courtesy of Ian Pickenpack, ATC Concepción)

Figure 9.69. LSQ airport terminal and ATC tower.

Figure 9.70. The maintenance and storage building is just behind the pine tree, the office spaces are in the bigger building next to it.

Figure 9.71. The wood frame was dislocated from the top beam.

Figure 9.72. Ropes were used to keep the wood frame wall from further damage.

Figure 9.73. A slightly damaged light fixture

Figure 9.74. Plaster walls cracks were observed inside the terminal building.

Figure 9.75. Terminal building with characters missing from the airport sign on the façade facing the airfield.

Air Traffic Control Tower

As the ATC room is at least two levels above the terminal building (Fig. 9.69), it sustained extensive damage. Almost all windows were broken. The damage started from the entrance of the staircase to the ATC room (Fig. 9.76). Large pieces of masonry collapsed at the first step. If the earthquake had occurred during operating hours, there would likely have been injuries. Most of the tower's glass panes were broken (Fig. 9.77). Unsecured equipment became dislodged and fell to the floor. The tower was evacuated and equipment was removed (Fig. 9.78). After the earthquake ATC operations were moved to the office space in the maintenance building immediately south of the terminal building.

According to the airport director, airport authorities from Santiago would perform an evaluation and develop a plan to restore the airport. However, she said that the airport was a low regional priority, and the assessment and repair effort be lower priorities as well.

Emergency Response

The airport was used for relief flights after the earthquake and was closed to commercial flights. A relief flight with medical supplies arrived during the team's visit on April 14.

Figure 9.76. Entrance to the ATC tower room; note the significant damage to the wall and the large pieces of masonry blocking the stair.

Figure 9.77. All the windows of this ATC tower were broken, except for one and a half panes.

Figure 9.78. Typical damage observed inside the ATC room.

9.6 Observations and Recommendations

The following represents of the observations of TCLEE team members during their visits as well as conversations with officials and they information provided. The recommendations are based on the observed failures; they are not meant to provide an analysis of the original designs, materials used, or construction practices.

1. SCL Passenger Terminal: Consideration should be given to improved connections for HVAC equipment and other items mounted on the ceiling and structural building elements. At the time of the TCLEE team visit, SCL technical staff members were evaluating alternative strategies for routing HVAC systems, including exterior mounting.
2. SCL Curbside: New supports should be designed and constructed for the pedestrian bridges from the terminal to the elevated curbside. Consideration should be given to less vulnerable evacuation routes from the terminal building.
3. SCL ATC Tower: New, seismically sound supports should be designed and installed for the ATC cab, with consideration given to base isolation to avoid damage due to shaking, which contributed to the loss of equipment in this case.
4. Concepcion (CCP): As this airport serves as the alternate to Santiago, consideration should be given to upgrading its emergency capacity, particularly its fuel supply, ATC capability, and radar installations.
5. CCP Passenger Terminal: Based on the failure experienced, a walk down of the all the nonstructural elements of the terminal building should be carried out in order to identify vulnerable areas and develop methods to improve seismic performance.

Runway Vulnerability and Survivability: The TCLEE team received inquiries about the possible vulnerability of the SCL runways to an event such as the Maule Earthquake. The availability of runways in the aftermath of a major earthquake is clearly an important issue. However, it is not a life-safety issue equivalent to the earthquake resistance of the passenger terminal. If a runway fails to survive an earthquake, air transportation is severely disrupted, potentially having a significant impact on the economy. However, if a passenger terminal collapses, the likelihood of personal injury and loss of life is extremely high. The following three aspects should be considered with the runways:

1. Survivability: The availability of a runway after a major earthquake will depend in large part on proximity to the epicenter and the underlying soil conditions. If constructed in a vulnerable location, especially on soils conducive to liquefaction, there is a high likelihood of damage severe enough to require runway closure. This was the case at Oakland in the aftermath of the Loma Prieta earthquake in 1989. Possible solutions include comprehensive soil improvement, the installation of stone columns to support runway pavement, or relocation of the runway altogether. None of these approaches offers a cost-effective solution.
2. Ease and Rapidness of Repair: Instead of making an existing runway resistant to earthquakes, it may be more cost effective to ensure that the airport has a plan to rapidly assess earthquake damage and make immediate repairs. This capability exists. The best example may be the U.S. Air Force "Red Horse" teams, which specialize in repairing bomb crater damage to runways so that they are quickly operational again.
3. Availability of alternative runways: As part of contingency planning at the airport, airline, and national levels, it is recommended that careful thought be given to the selection and preparation of alternate runways outside of the vulnerable area.

9.7 Acknowledgments

The authors are indebted to the individuals listed below for their kind support providing access to critical facilities within the airport and providing photographs showing earthquake damage for this report as well as permitting us to use those selected for this chapter.

- Mr. Guillermo Guzman Solar, Director Airports Projects, Ministry of Public Works,
- Mr. Eduardo del Canto Hidalgo, Civil Aeronautics Administration (DGAC),
- Mr. Rodrigo Lopez Muela Parra, Civil Aeronautics Administration (DGAC),
- Mr. Juan Pablo Muñoz Díaz, LAN Chile Airlines,
- Mr. Hector Sandoval, Civil Aeronautics Administration (DGAC), Concepción Airport
- Mr. Rodrigo Esquivet Opazo, Asesoria Inspección Fiscal Aeropuerto Carriel Sur Acting Administrator,
- Ms. Alejandra Ruiz, Civil Aeronautics Administration (DGAC), Los Angeles Airport Administrator.

10 INFRASTRUCTURE INTERDEPENDENCIES AND RESILIENCE

The February 27, 2010 Mw 8.8 offshore Maule, Chile earthquake revealed aspects of infrastructure interdependence that increased the loss of functionality of lifeline systems already affected by direct earthquake damage. Additionally, interdependencies delayed the restoration processes of certain susceptible and dependent utility systems, such as telecommunication networks, power distribution grids, water networks, and transportation systems. The observed loss of functionality and delayed restoration times from interdependence also contributed to reduced resilience—a central concept in infrastructure engineering related to the time to restore utility service to pre-event levels. In general, interdependencies were triggered by physical and cyber interactions between lifeline systems as well as from specific cases of collocation and from relational and logistical coupling between infrastructures and institutional entities. The later kinds of interdependencies were particularly undesirable between telecommunication systems and emergency evacuation institutions during the tsunami triggered by the earthquake. Time series analyses of restoration curve cross-correlations are also introduced to quantify interdependencies and discuss their values in light of anecdotal accounts from the field. This chapter concludes with a series of research and practical questions to be addressed by the lifeline earthquake engineering community.

10.1 Phases of Lifeline System Interdependence

Physical and cyber coupling led to loss of lifeline functionality that manifested in two main phases. Phase I was on average a short period phase filled with social turmoil, power outages, and high uncertainty about tsunami evacuation and road network conditions. Specifically, this phase was characterized by blackouts at the power transmission (> 66 kV) and sub-transmission (13.3 to 23 kV) levels, which affected the telecommunication systems, mostly mobile communications because of the nearly nonexistent emergency power at cell sites beyond the 3 to 4 hours of battery backup provided on average. In turn, telecommunication system outages affected the undamaged portions of the power distribution infrastructure as damage and safety assessments of its components were delayed. The water infrastructure had adequate emergency power to operate treatment plants, but the direct earthquake damage to the water mains and laterals prevented operation system. In addition, the uncertainty on fuel availability, refinery shut downs, and road conditions hampered the operation of the undamaged and repaired portions of the water systems network during the power and telecommunication outages. Additionally, the telecommunications outages delayed some of the early hiring, coordination, and assignment of water system crews to field repairs. This did not change the number of affected customers—only the time to restore service. The gas network most likely had a similar performance issues due to pipeline susceptibility to ground failure. Although Phase I durations differed by region, it lasted for approximately 3 days in the Concepción area.

Phase II was characterized by the increasing availability of alternate transportation routes, restoration of power at the transmission and sub-transmission levels, and regained social order. This phase also evidenced the emergence of physical system and institutional interdependencies that exacerbated restoration delays at focused locations. These observations enabled a steady recovery of telecommunication systems, although significant congestion was experienced at first. Additionally, repair of water, gas, and transportation systems seemed to be more efficient and better coordinated, although it was difficult to achieve 100 percent restoration. In fact, an improving restoration of feeders and laterals but with a decreasing restoration rate overall was the norm in this phase for power distribution systems and with most other utility systems. Unique circumstances contributing to the decreasing restoration rate included sites with access limitations, the infeasibility of maintaining diesel generators for communication or construction equipment, interdependencies, and sites under special constraints, such as those with no utility repair orders until demolition of unsafe buildings occurred. These trends were affected by the reduced consumption base caused by loss of residential and industrial utility users, mainly in coastal areas affected by the tsunami.

The steady infrastructure performance recovery during phase II also led to decreases in physical interdependence while promoting relational interdependence at site-specific locations. This relational interdependence occurred between lifeline systems and small yet representative groups of citizens, government, and service institutions and contributed to indirect losses. The interplay between high and widespread physical/cyber interdependence effects in phase I and high but focused relational interdependence along with low and focused physical/cyber interdependence phase II has been observed in other disasters, including intentional attacks, ice storms, and hurricanes (Mendonca and Wallace 2006, Chang et al. 2009). These non-seismic studies also point out that the proportion of interdependence-induced incidents account up to 25 percent of the post-event reported consequences—a significant percentage but certainly not the sole factor contributing to lifeline system functionality loss and restoration delays.

The most common instances of relational interdependence effects involving human, institutional, and lifeline system coupling were adequately managed by most lifeline systems companies. Thanks to reliably available workers, supported by food provisioning and relatives search programs, these companies were able to avoid operational personnel deficiencies. However, institutional coupling had some detrimental effects. Some service companies, such as banks in dense urban areas, had idle time. Despite having completed internal retrofit projects, they were not scheduled to receive power and other utility services until all tagged buildings had been demolished. The demolition and reinstallation of utility infrastructure interrupted business for another 3 to 4 months. In addition, as the restoration rate of most lifeline systems slowed after the majority of customers came back on line, the remaining residential and commercial users endured significant inconvenience and indirect losses from direct dependence on lifeline systems, delayed recovery of the local economy, and disruptions to international trade. This physical-societal interdependence was prevalent during phase II and greatly affected the resilience of earthquake-damaged areas.

Lifeline system damage mechanisms beyond direct damage and physical or relational interdependence included geographical collocation. Despite important underground, overhead, and surface levels of collocation cases of infrastructure system being observed in the field, only a few materialized into interdependence-induced failures. These specific cases included telecommunication, gas, and water lines conveyed by collapsed or excessively displaced bridges at waterway crossings, electric train halts from power and telecommunication downed poles, rooftop telecommunication structure failures or cut operations at collapsed or tagged for demolition buildings, and power distribution overhead lines pulled down at tapping points from collapsed housing facades or entire structures (Fig. 10.1 and 10.2).

Figure 10.1. Surface collocation of telecommunication and power lines conveyed by bridges of the road network. This event was registered at the Juan Pablo II bridge over the Biobío River in Concepción.

Figure 10.2. Surface collocation of power delivery and electric train systems between Los Angeles and Linares. Telecommunication lines are also conveyed by the power poles. Poles were susceptible to slope instabilities and excessive top end displacement triggered by pole-mounted weights used to keep tension on power lines.

10.2 Lifeline System Restoration Performance

Lifeline system restoration data was obtained from the field investigations by the TCLEE team, the Ministry of the Interior of the Chilean government, and reports from utility companies in Regions VI through VIII, including the water and wastewater company ESSBIO and the power transmission and distribution company, CGE. Specifically, data on the percentage of service availability at the distribution facilities or substations as a function of time was obtained for power and water distribution in the cities of Concepción and Talcahuano, which mostly represent inland and coastal communities in Region VIII, respectively. Additionally, average restoration curves for the entire Regions VII and VIII were obtained for power and telecommunications (fixed and mobile telephones). Power restoration data was also collected for the cities of Talca and Constitución, representing inland and coastal communities in Region VII, respectively. Note that the percentage of customers restored is widely used as a recovery metric in practice, but in fact, utility operators have little information on how many customers have actually been restored. As a particular example, when power systems' operators talk about customers restored in the distribution system they are typically referring to the number of customers on the feeders that have restored power. In practice, it is still difficult to know how many customers actually have power, as there can be damage to the distribution lines once they leave the distribution substation. Smart grid systems may address these issues in the future. Other utility systems also tend to report potential customers associated with their restored service area distribution units, not actual customers with restored service—although these values are in general positively correlated.

Figure 10.3 presents the available average data on lifeline system recovery for Region VIII. Note the different global trends between water system recovery and power or telecommunications. There is a higher availability of power across Region VIII than mobile phone service in the same area. Power restoration in Concepción and Talcahuano also tends to be higher relative to mobile telecommunication systems in the entire Region VIII, except for a period at the end of the first week and start of the second week. These trends suggest a tight functional interaction between power and communication systems. Note the higher availability of fixed telecommunication landlines relative to mobile phones over the entire restoration phase, a fact exploited by most utility companies to organize their logistics via fixed telephone access. This is because in practice, central offices with more than 5000 telephone subscribers typically had adequate backup power generation, and although they were affected by congestion, most of them never lost service. Finally, the evident direct seismic damage to the water distribution system overshadowed interdependent effects on operation or restoration from power, telecommunications, or transportation of fuel.

As lifeline restoration is in practice initially performed without completely adhering to reliability standards or knowing the sources of damage, detailed inspections of surviving equipment tend to slow down the restoration process. This is reflected in the decreasing restoration rates as service restoration increases but at a slow pace. As an example, power systems have a target of 3.5 interruptions and 13 hours of interruption per year in urban areas, and 5 interruptions and 15 hours of interruption in rural areas. The unserved customers during the power restoration process of phase II (including power delivery, water, and gas lifeline operators) faced congested telecommunication systems and fuel shortages for emergency power generation due to difficulties in transporting gasoline and diesel through abnormally long routes within the available road system, along with the shutdown of important refineries in the Concepción area. This phenomenon highlighted the potential additional effects of interdependences from transportation systems and supply chain logistics (including other lifeline systems), which mainly affected the length of infrastructure system restoration times and overall resilience through degraded restoration rates.

Figure 10.4 presents the existing service restoration curves for Region VII. The restoration trends for power and telecommunication systems are remarkably similar to the recovery in Region VIII, although recovery in Region VII was slightly faster. The figure also shows a wide gap between the restoration of coastal (Constitución) and inland (Talca) cities given the severe damage in the tsunami-affected area. This gap is not observed in Region VIII because Concepción and Talcahuano are geographically adjacent and tsunami effects were not as prominent as in Constitución. Additionally, Constitución showed a higher percentage of collapsed or severely damaged buildings than in the other cities due to its closer proximity to the earthquake epicenter and by the higher percentage of old adobe buildings. The extensive damage and simultaneous permanent loss of customers in Constitución also played a role in making its restoration slightly faster than the coastal Talcahuano city in Region VIII.

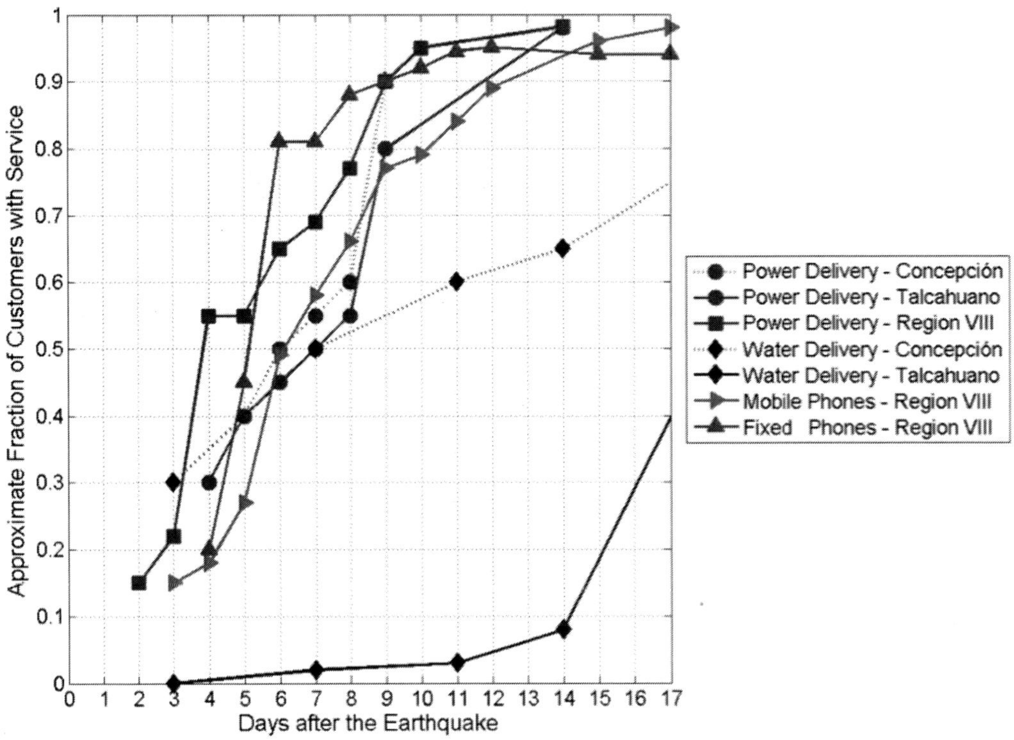

Figure 10.3. Service restoration time series for available utility systems in select cities and across the Biobío Region VIII. Note that the fraction of customers is an approximate value as the actual measurement of recovery is typically the availability of distribution facilities or substations in charge of specific service areas.

10.3 Quantitative Lifeline Interdependence Analysis

The time evolving nature of the restoration curves allows a natural exploration of the data in time series contexts. These analyses permit quantification of the leading or lagging relations in the data, reveal hidden restoration trends, and provide an alternative way to measure interdependence strength across lifeline systems. Details on the time-series treatment of restoration curves are presented in Duenas-Osorio and Kwasinski (2011) where data de-trending and other transformations are necessary to apply standard time-series analysis tools. Emphasis is placed on the cross-correlation function (CCF) across restoration curves, which is a dimensionless metric of the cross-covariance function and the vehicle to analyze the joint behavior between different time series and quantify interdependence. Typically, the sample cross-correlations are examined graphically as a function of the lag time h to search for leading or lagging relations, as well as for comparison of significant peaks to reveal coupling strengths. All the results from time series analyses are discussed in the context of anecdotal information collected by the field investigation team to properly interpret cross-correlation values and justly impute causality.

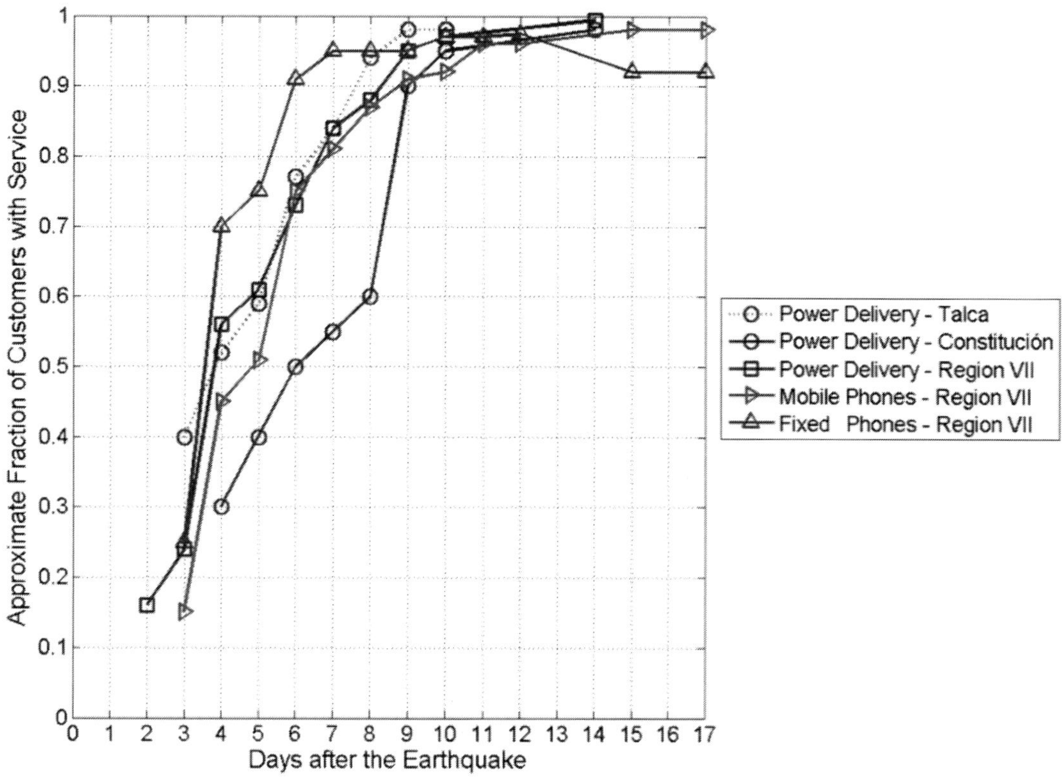

Figure 10.4. Service restoration time series for available utility systems in select cities and across the Maule Region VII. Note that the fraction of customers is an approximate value as the actual measurement of recovery is typically the availability of distribution facilities or substations in charge of specific service areas.

Plots of CCFs versus lag times are presented in Appendix Figures A1-A21 underneath the detrended series for all restoration curve pairs associated within the hardest hit Region VIII. As an illustration, Figure A2 shows two restoration curves, highlighting a case with high correlation at a low positive lag time of $h = 2$, in which the positive sign indicates that the fixed phone restoration lags relative to the power restoration. This relation was only hinted in Figure 10.3 but confirmed and quantified in the CCF plot. The small value of h further confirms that the two systems affect each other's operation. In contrast, Figure A17 shows two restoration curves with weak correlation at a high negative lag time of $h = -8$, where the negative sign indicates that power delivery in Concepción has a tendency to lead, albeit indirectly, the restoration of water delivery. The large negative h compounds the observed weak correlations to reveal that restorations in power systems at the local level have little direct effect on the service restoration of water delivery systems due to the heavy damage of water pipelines. However, cross-correlation effects at high lag or lead levels may not indicate instantaneous operational coupling but rather delayed logistical coupling (i.e., water utility repair crews and equipment coordination once power was restored).

Table 10.1. Symmetric Cross-Correlation and Lag Values for Restoration Curves in Region VIII.

Series	Fixed Telephone		Mobile Telephone		Regional Power		Power Concepción		Power Talcahuano		Water Concepción		Water Talcahuano	
	Peak ρ	Lag h	Peak ρ	Lag h	Peak ρ	Lag h	Peak ρ	Lag h	Peak ρ	Lag h	Peak ρ	Lag h	Peak ρ	Lag h
Fixed Telephone	1.00	0.00	0.74	0.00	0.84	2.00	0.53	2.00	0.74	-3.00	0.66	-11.00	0.96	-11.00
Mobile Telephone	0.74	0.00	1.00	0.00	0.73	2.00	0.64	2.00	0.83	-3.00	0.48	-11.00	0.74	-11.00
Regional Power	0.84	-2.00	0.73	-2.00	1.00	0.00	0.79	0.00	0.89	-5.00	0.56	-13.00	0.79	-13.00
Power Concepción	0.53	-2.00	0.64	-2.00	0.79	0.00	1.00	0.00	0.68	-5.00	0.35	-8.00	0.53	-8.00
Power Talcahuano	0.74	3.00	0.83	3.00	0.89	5.00	0.68	5.00	1.00	0.00	0.50	-8.00	0.75	-8.00
Water Concepción	0.66	11.00	0.48	11.00	0.56	13.00	0.35	8.00	0.50	8.00	1.00	0.00	0.70	0.00
Water Talcahuano	0.96	11.00	0.74	11.00	0.79	13.00	0.53	8.00	0.75	8.00	0.70	0.00	1.00	0.00

To explore more fully the information from cross-correlations, Table 10.1 presents a summary of the peak CCF values represented by ρ for all restoration curve pairs in Region VIII along with their lag values h. The table shows that as with fixed phones, mobile phones also have a high correlation with regional power delivery at $h = 2$, indicating strong operational interdependence. Interestingly, regional power and telecommunications, along with local power, have a high correlation with water delivery but with high negative lag times, confirming indirect water system dependence on other utilities. This dependence is certainly not operational due to the high h but rather logistical in terms of facilitating coordination of repairs. These repairs only manifest as water service restoration after more than a week from the earthquake. Fixed and mobile phones also exhibit high correlation at no lag time, as does the water in adjacent inland and coastal cities. Power in Concepción leads the recovery of power in Talcahuano by almost a week, possibly due to the tsunami effects in the coastal city. Power at the regional level leads coastal city restorations or interacts instantaneously with inland city restorations.

10.4 Research and Practical Challenges

The site observations, professional discussions, and field data on lifeline system interdependencies and resilience along with the quantification of their coupling strengths triggered several questions for the lifeline earthquake engineering community. These questions, relevant to Chile as well as the United States and other countries, include:

1. **How can risk-consistent national-level design provisions be established for dissimilar and interdependent lifeline systems?** There is a strong sentiment to pursue the development, dissemination, and adoption of guidelines and subsequent standards for the design and maintenance of utility systems within a risk/resilience framework. Risk-based philosophies require studying the lifeline systems at the

transmission, distribution, and institutional levels. Exiting guidelines have disparate design goals based on reliability, cost, risk, operation, and such.

2. **What performance goals are needed to retain design consistency across utility systems but at the same be informative to the different stakeholders of the distinct lifeline systems?** Performance expectations and their measurement vary according to the stakeholders as their interests can range from quantification of service disruption, system tolerance, and restoration times, to reimbursement projects costs and customer satisfaction. Performance goals also change from either planning to post-event scenarios. Hence, regardless of these multiple requirements of performance, a consistent risk/resilience target should guide the control and management of lifeline system performance.

3. **How can the current availability of vulnerability models be increased for lifeline systems at both the component and system levels, and how can these models relate to loses?** It became clear that although expert-based assessments of vulnerability are valuable, there is a pressing need for computational, theoretical, and empirical developments of vulnerability models for scenario and aftershock analyses. In addition, the assessment of indirect impacts remains a challenge due to its interdisciplinary nature, although it also provides an opportunity to clearly link society to lifeline systems engineering and effectively quantify resilience.

4. **Can new protocols to collect lifeline system performance data be established?** Databases with lifeline system performance during normal and abnormal operation are not yet available to validate emerging models and techniques for complex network analysis and interdependent effect evaluations. Performance data is essential as well as anecdotal accounts about resources and logistics for restoration and maintenance processes. Causal and correlation effects are necessary for interdependency modeling as well as processes for interconnection between different companies and across jurisdictions.

5. **How do the emerging interdependent system modeling techniques relate to one another?** About 75 percent of the literature on infrastructure system interdependencies is very recent (within 5 years); thus, there is no clear understanding of how techniques based on input-output models, agent-based models, probabilistic network-based approaches, time series analyses, and empirical formulations relate to one another. It is unclear if the techniques are complementary or competing at this early stage of intellectual contributions.

6. **Can interdependency analysis models go beyond physical coupling?** Economic, institutional, and environmental forces on the existing models of interdependence are lacking. An adaptive systems approach is required if models to assist the planning and management of complex lifeline systems are to approach reality conditions, including climate change, decision-making, budgetary constraints, and such. Theoretical and computational techniques are yet to be developed to address this multi-scale coupling and resilience.

7. **Has the uncertainty in lifeline system design, operation, and response been properly quantified?** To effectively communicate risk to utility owners and users, uncertainty source identification and quantification is essential for credibility and a

building block for the development of guidelines and standards. In addition, the taxonomy of uncertainty can guide the use of resources to effectively reduce or control uncertain performance outcomes.

8. **Can experimental work help reduce the uncertainty in lifeline system modeling and future design?** To achieve risk-based homogeneous performance of distinct utility systems, some particularly vulnerable failure mechanisms need to be explored in detail. For instance, buried pipes for water/wastewater and gas are consistently overly stressed by ground failure during ground shaking events, but no systematic information from experimental designs exists to fully characterize the failure rates and the factors that influence their response the most. Other examples of tests include elements of the power grid from both legacy and modern equipment, telecommunication network components, and the performance of distinct operational regimens. In addition, temporary and mobile utility system units, such as water and telecommunication systems, are yet to be evaluated under extreme hazard and aftershock conditions. Alternative materials such as HPDE for pipes and anchoring techniques for racks and trays of cables and equipment need experimental certification.

10.5 Conclusions and Recommendations

Observations and results from this reconnaissance work show that interdependencies are not negligible. In fact, a key insight from cross-correlation analyses of restoration time series along with their lag or lead time effects includes the relevance of telephone systems both fixed and mobile to their own operation and to the operation of power delivery systems and vice versa. Intradependence is also observed, referring to couplings within systems of the same kind but studied at different jurisdictional sizes, particularly when regional restorations led local recovery or inland cities led coastal restoration. In addition, at the city level, power delivery shows important correlation with the restoration of water systems, a fact not clearly conveyed by the raw restoration curves alone. Other utilities at the regional level also show important correlation with water systems; however, such correlation is deemed as a logistical not operational coupling given the significant lag times of more than one week.

Regarding lessons learned and strategies that reduced detrimental effects from interdependencies and enhanced resilience, water companies allocated all fuel from wastewater treatment plants for the potable water system, installed portable tanks to supply water in a decentralized fashion, and relied on commercial radio stations to coordinate contractors and repair crews. In addition, power delivery systems with underground cables in downtown areas (away from tsunami-impacted sites) reported no damage and were able to provide electricity upon restoration of the transmission system; however, these underground cables only represent a small percentage of the delivery network. Gas companies adopted individual cylinder delivery approaches instead of reconstruction of piping systems, and coal-fired power plants in coastal areas used available salt water for cooling as standard practice. Lifeline system operators also adopted radio (power delivery and water systems), satellite phones (power delivery systems), and available landlines to enhance their collapsed mobile communications within and across departments of their organizational structures. At the infrastructure and institution interface, mobile schools were installed during the third week after the earthquake to help restore normalcy and resilience to the community. These schools were located in areas were new power delivery infrastructure was installed, highlighting another instance of long-term interdependence at the intersection of physical and social systems.

Overall, the reported logistics of restoration and performance of individual systems have room for improvement to reduce periods where interdependent lifeline systems are unavailable. Better performance of the transportation system for post-earthquake mobilization of fuel and supplies, faster damage detection at the power delivery system, and alternate power along with congestion management of the telecommunication systems could have reduced the overall impact of the earthquake during the power transmission and early power delivery blackout periods and reduced the deadly consequences of the tsunami. The long-term lifeline system repairs post power restoration indicate the potential benefits of direct damage prevention through soil improvement; higher seismic design standards at critical locations for bridges, water, and gas systems; the availability of alternate power delivery feeders for network reconfiguration; the reduction of collocated lifeline points; and the existence and training of utility personnel according to seismic risk management plans. More specifically, diverse power supply at communication sites through distributed generation may reduce power outages. In cell sites with low power consumption, the use of alternative power sources, such as photovoltaic systems and wind generators is a very effective solution to decouple lifeline systems. Call congestion issues can be addressed by a call priority program such as that implemented in the United States by the National Communications System. In addition, the adequate performance of transportation networks, along with their redundancy and availability of short detours at critical locations can reduce the restoration times of lifeline systems and decrease the effects of socio-technical interdependent effects. The upgrade of power line communications for integrated Supervisory Control and Data Acquisition (SCADA) also can improve power transmission system coordination across multiple providers and perhaps recovery times, especially if extended to power distribution systems. Finally, the use of high-density polyethylene (HDPE) in certain main water distribution pipes demonstrated outstanding performance under earthquake loads at the limited locations where it was used, which can reduce the lengthy restoration times of buried pipelines.

10.6 Acknowledgements

The TCLEE team is grateful with the logistical and technical support from the Universidad de la Santísima Concepción in Chile through Professor Mauricio Villagrán, and the Pontificia Universidad Católica de Chile through Professor Hugh Rudnick. The TCLEE team also acknowledges the support and data sharing from ESSBIO, the water and wastewater company in Regions VII and VIII in Chile, and the Compañia General de Electricidad (CGE).

10.7 References

Chang, S., McDaniels, T., and Beaubien, C., 2009. Societal impacts of infrastructure failure interdependencies: building an empirical knowledge based, Proceedings of the 2009 Technical Council on Lifeline Earthquake Engineering (TCLEE) Conference, Oakland, CA, June 28-July 1, 2009, p.p. 693-702.

Dueñas-Osorio, L., and A. Kwasinski, (2011). " Quantification of Lifeline System Interdependencies after the 27 February 2010 Mw 8.8 Offshore Maule, Chile Earthquake." Earthquake Spectra, in review.

Mendonca, D., and Wallace, W. A., 2006. Impacts of the 2001 World Trade Center attack on New York city critical infrastructures, ASCE Journal of Infrastructure Systems 12(4), 260-270.

10.8 Appendix A

Pairwise plots of detrended restoration curves and cross-correlations are included for available utility systems and jurisdiction sizes in Region VIII. The top system in the legend leads the performance of the companion system for maximum positive CCF values in the negative side of the lag times. For maximum positive CCF values in the positive side of the lag times, the top system lags in performance relative to the companion system. Peak values and associated lag times are summarized in Table 10.1. The horizontal lines in the CCF plots indicate the significance of the peaks in large samples.

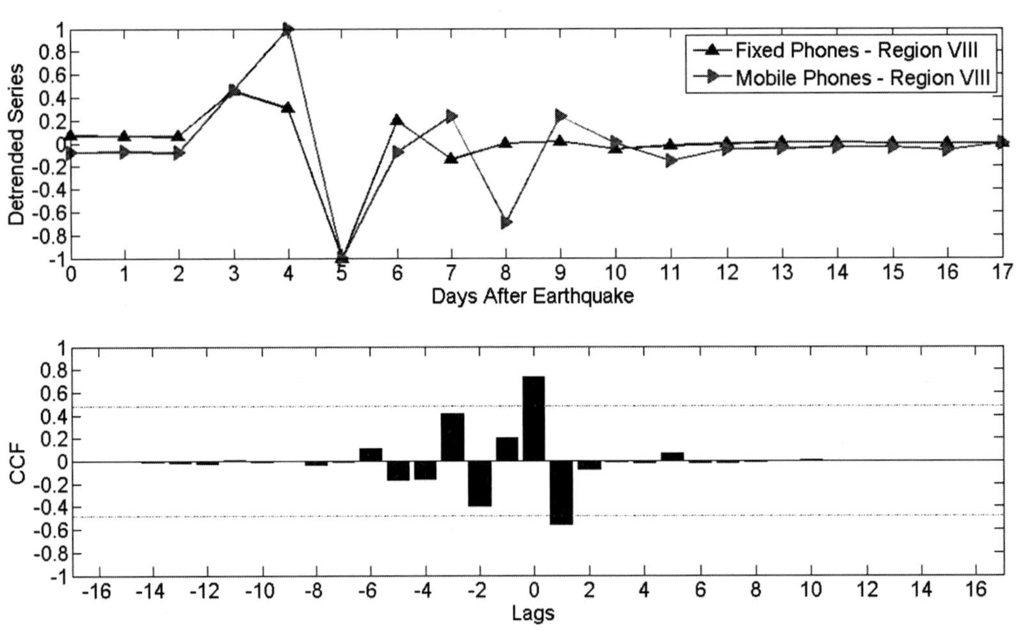

Figure A.1. Detrended series and cross-correlation CCF values between fixed phones and mobile phones.

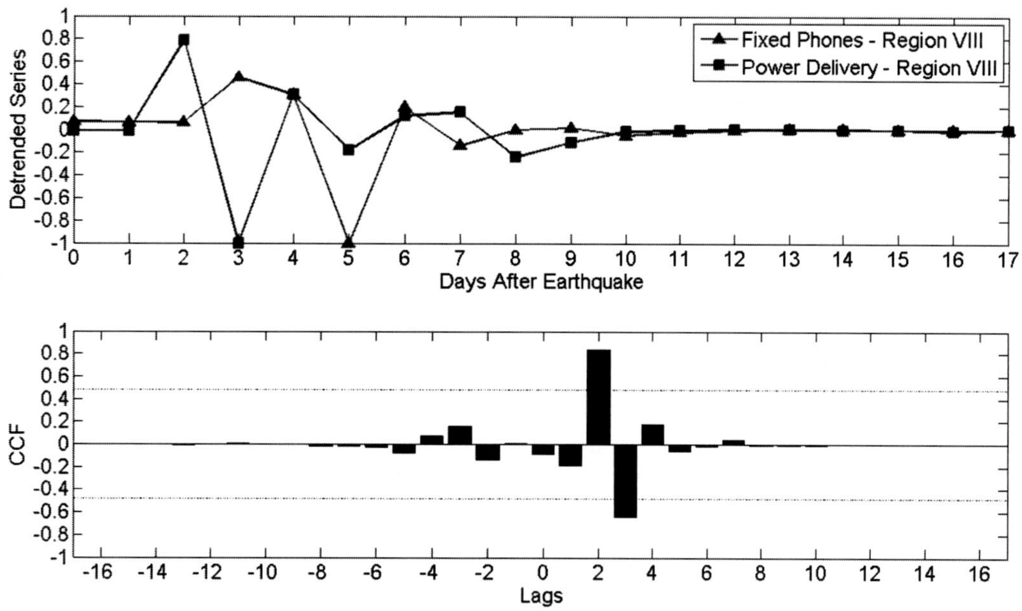

Figure A.2. Detrended series and cross-correlation CCF values between fixed phones and regional power.

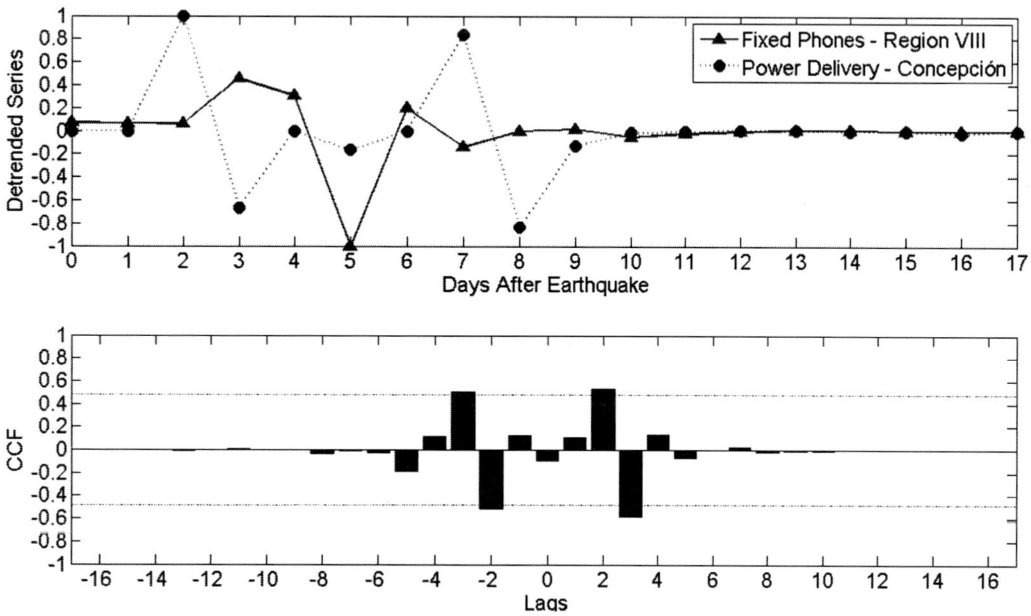

Figure A.3. Detrended series and cross-correlation CCF values between fixed phones and power in Concepción.

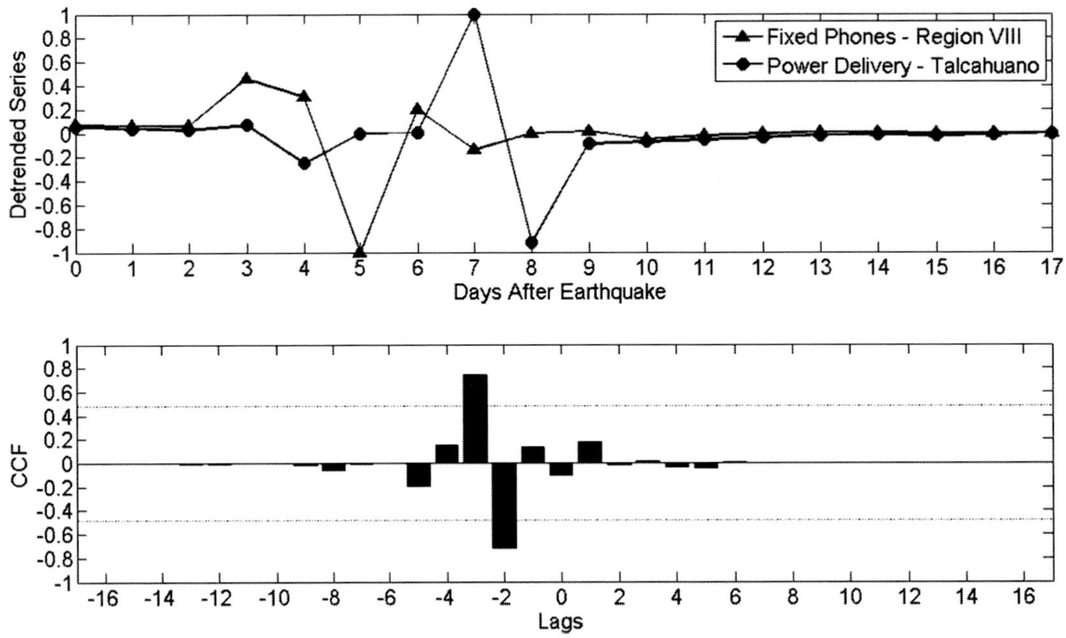

Figure A.4. Detrended series and cross-correlation CCF values between fixed phones and power in Talcahuano.

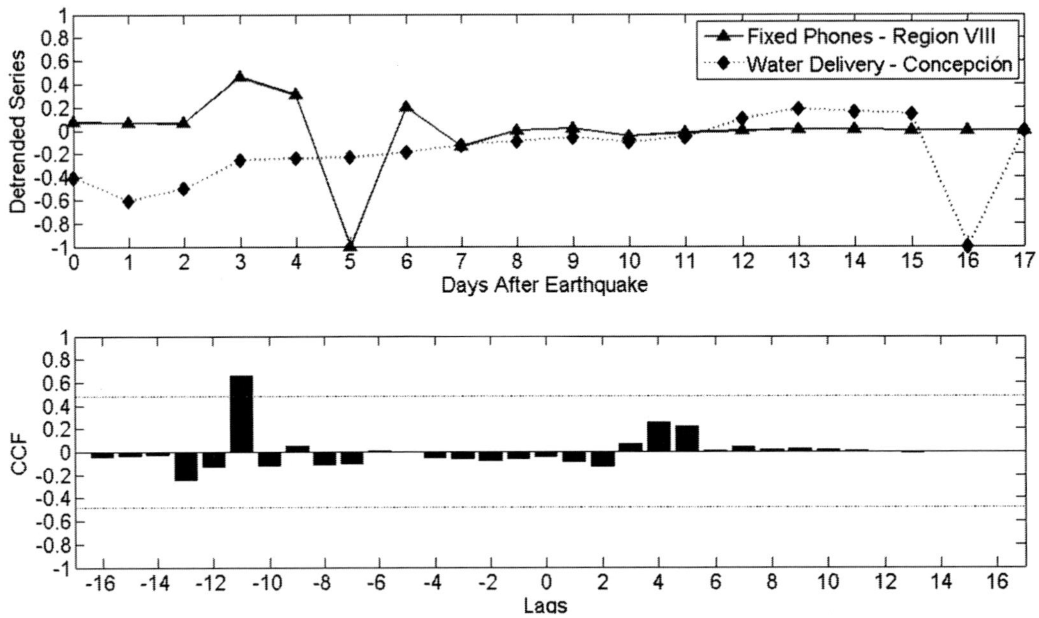

Figure A.5. Detrended series and cross-correlation CCF values between fixed phones and water in Concepción.

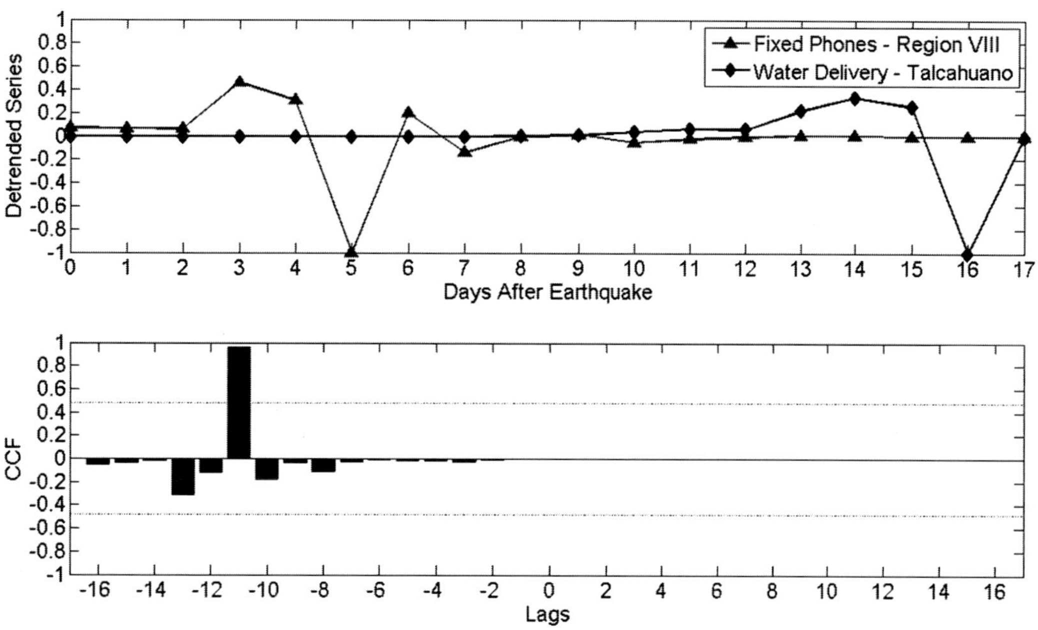

Figure A.6. Detrended series and cross-correlation CCF values between fixed phones and water in Talcahuano.

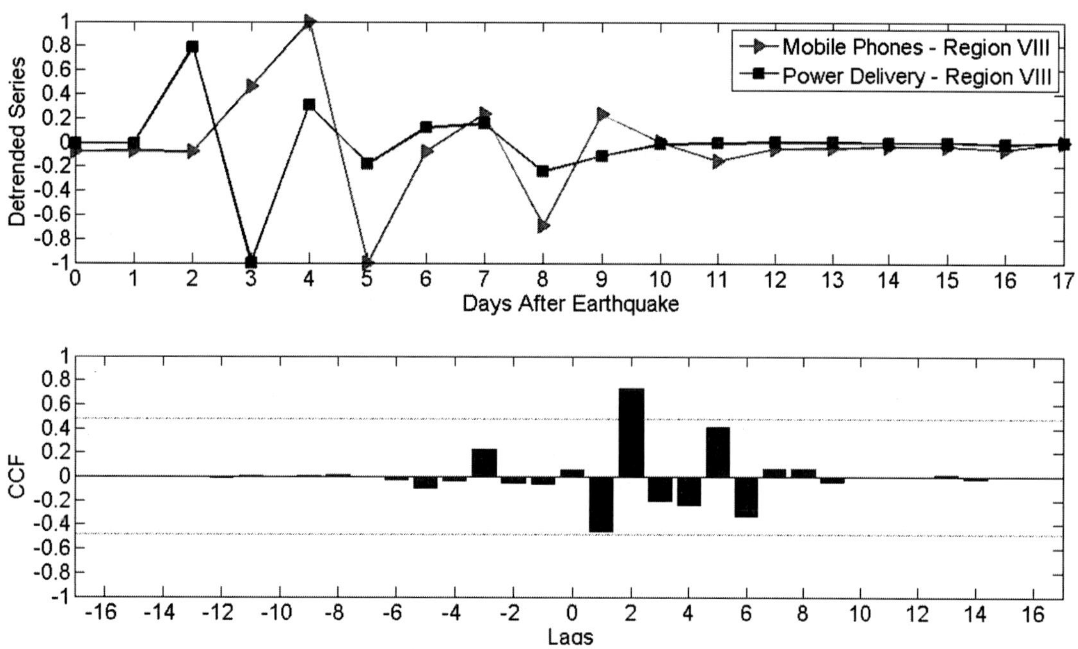

Figure A.7. Detrended series and cross-correlation CCF values between mobile phones and regional power.

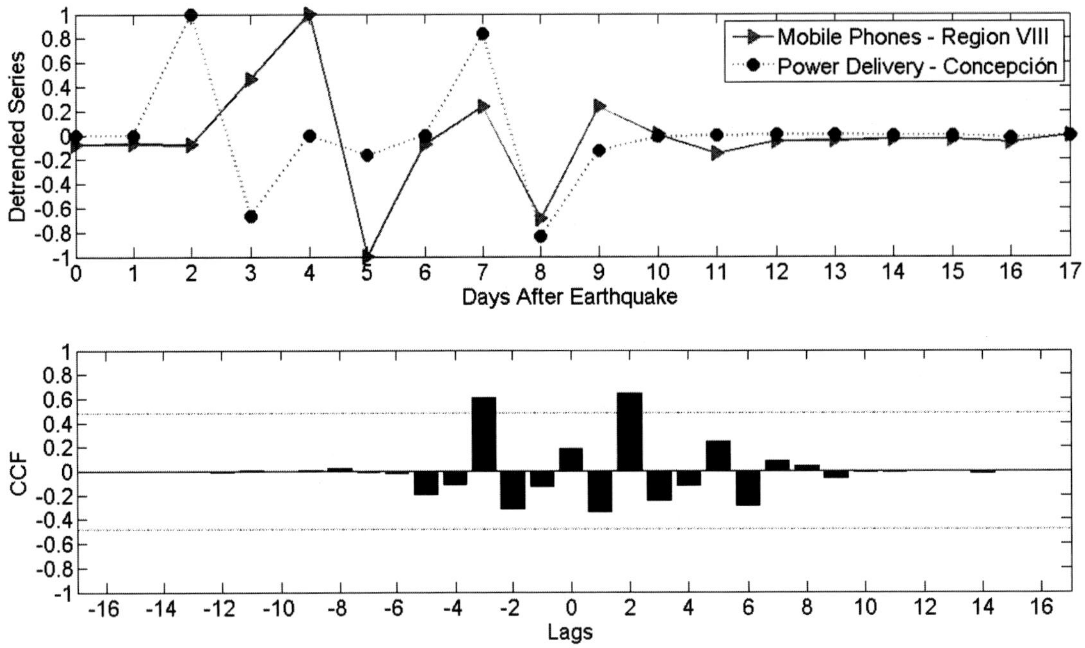

Figure A.8. Detrended series and cross-correlation CCF values between mobile phones and power in Concepción.

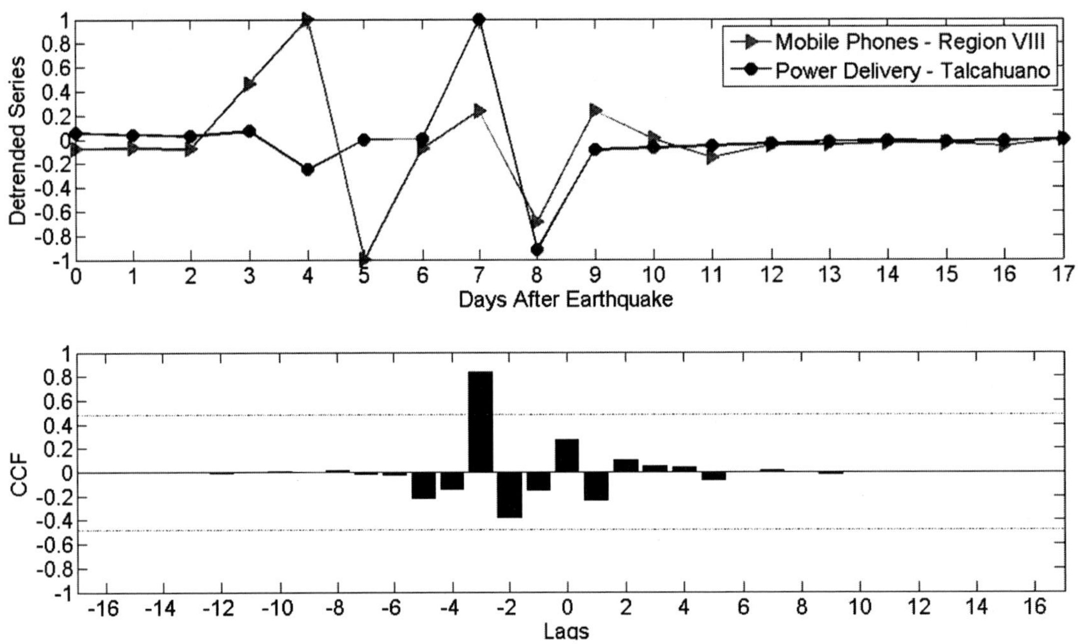

Figure A.9. Detrended series and cross-correlation CCF values between mobile phones and power in Talcahuano.

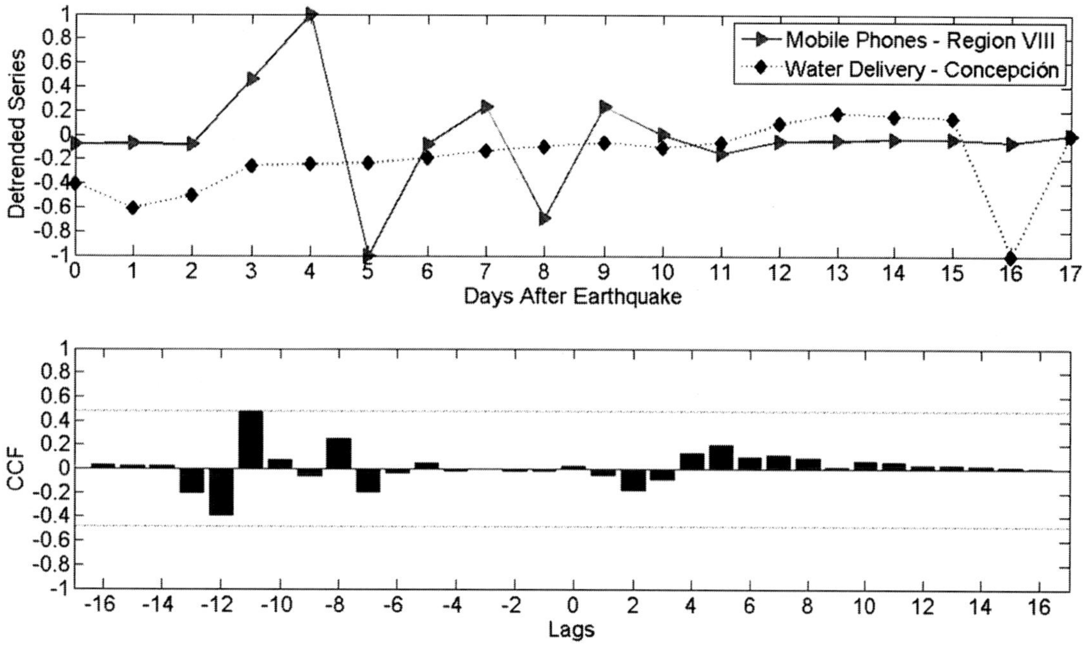

Figure A.10. Detrended series and cross-correlation CCF values between mobile phones and water in Concepción.

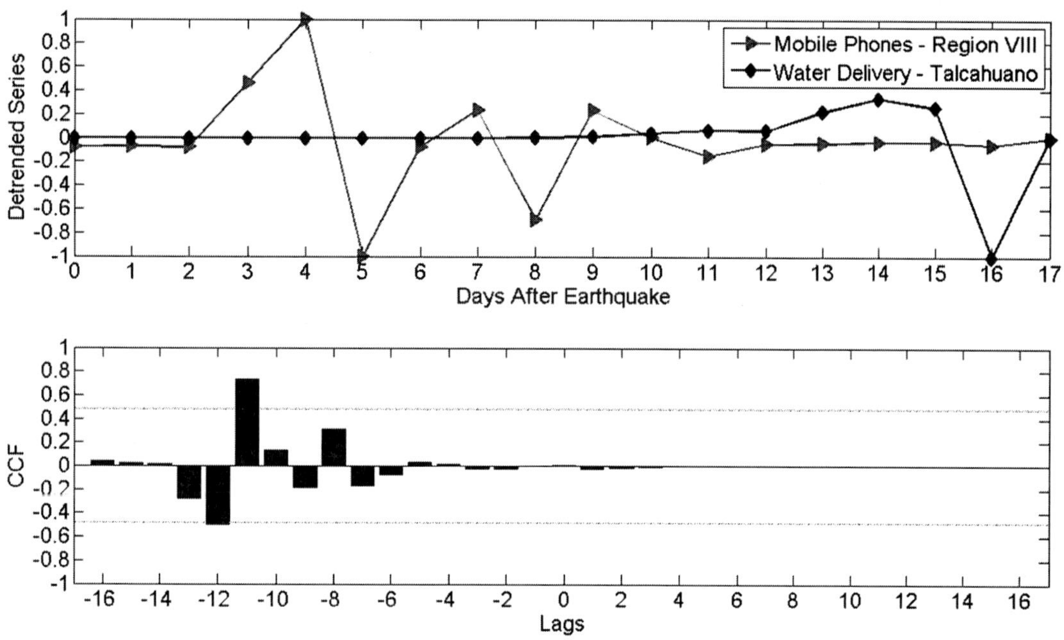

Figure A.11. Detrended series and cross-correlation CCF values between mobile phones and water in Talcahuano.

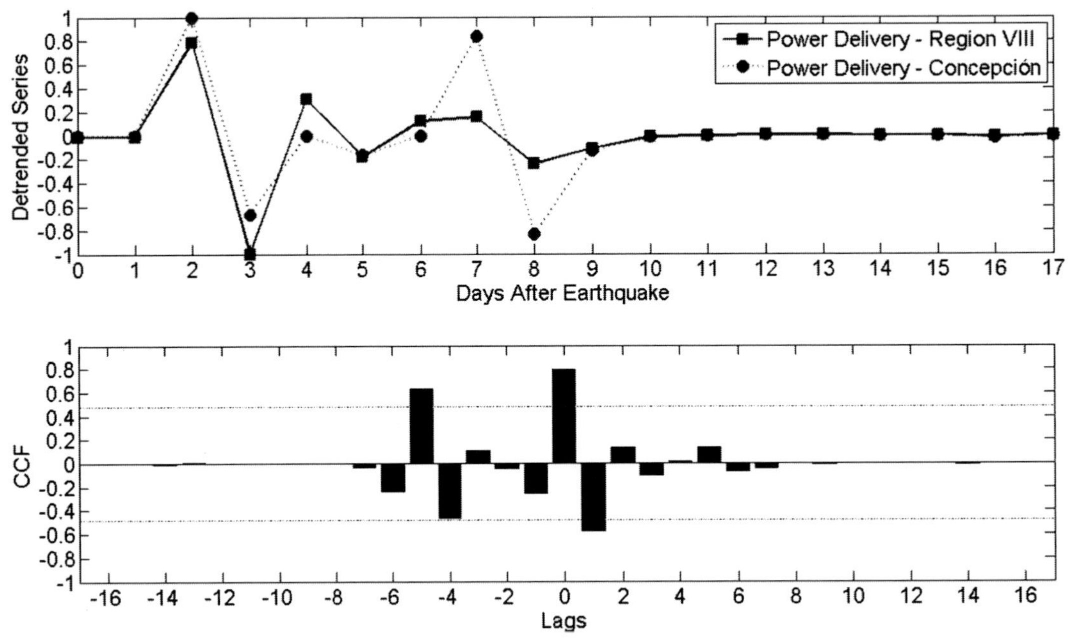

Figure A.12. Detrended series and cross-correlation CCF values between regional power and power in Concepción.

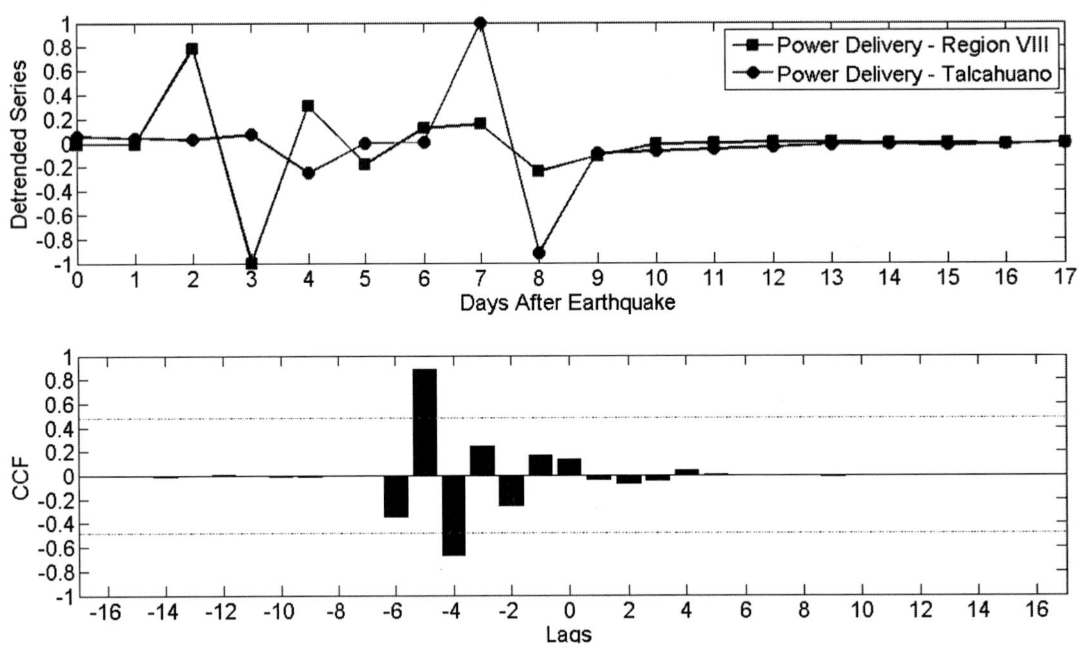

Figure A.13. Detrended series and cross-correlation CCF values between regional power and power in Talcahuano.

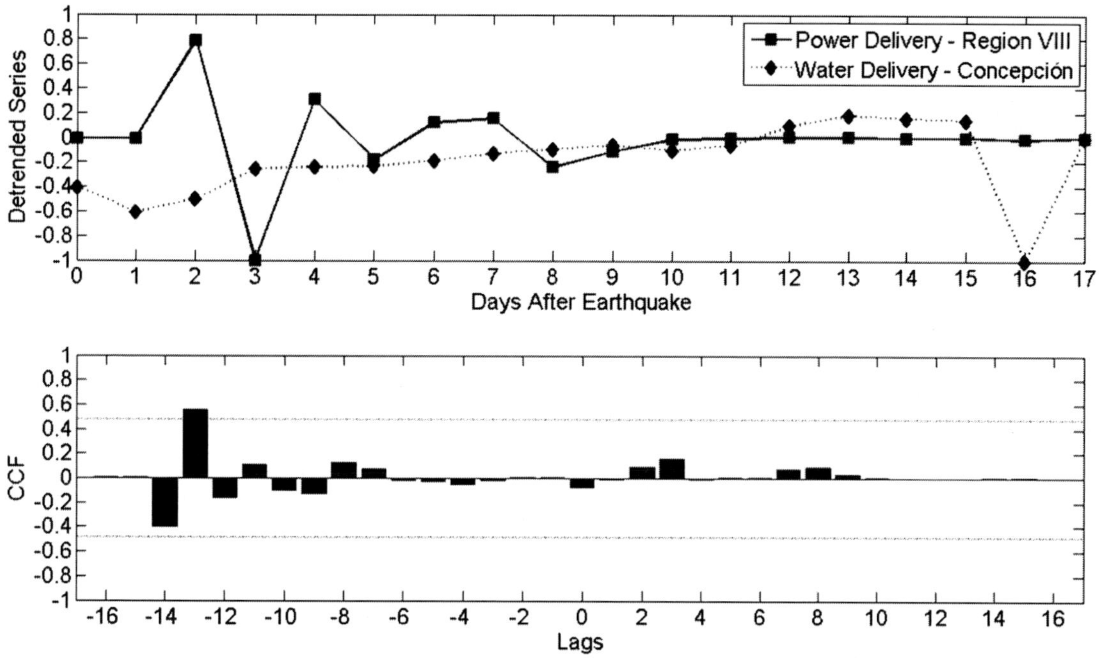

Figure A.14. Detrended series and cross-correlation CCF values between regional power and water in Concepción.

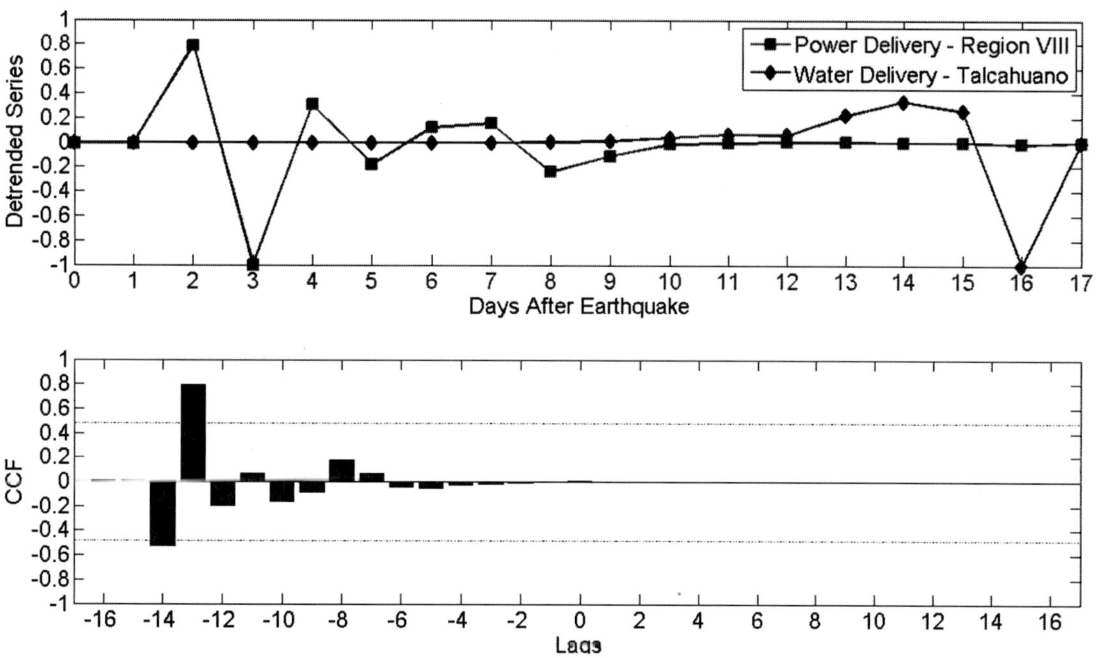

Figure A.15. Detrended series and cross-correlation CCF values between regional power and water in Talcahuano.

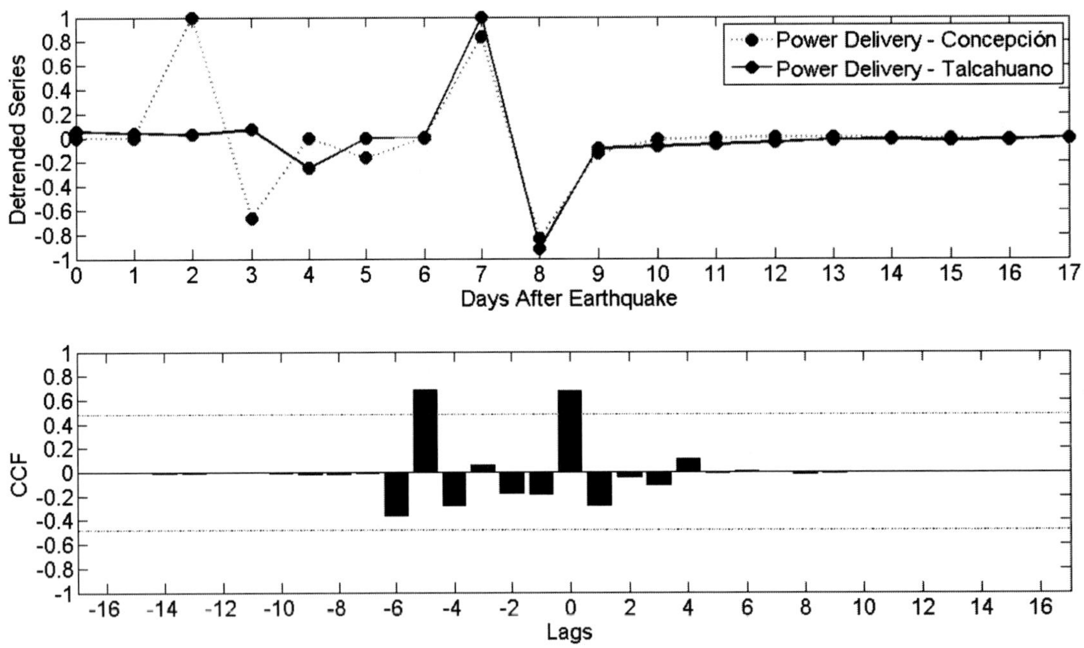

Figure A.16. Detrended series and cross-correlation CCF values between power in Concepción and power in Talcahuano.

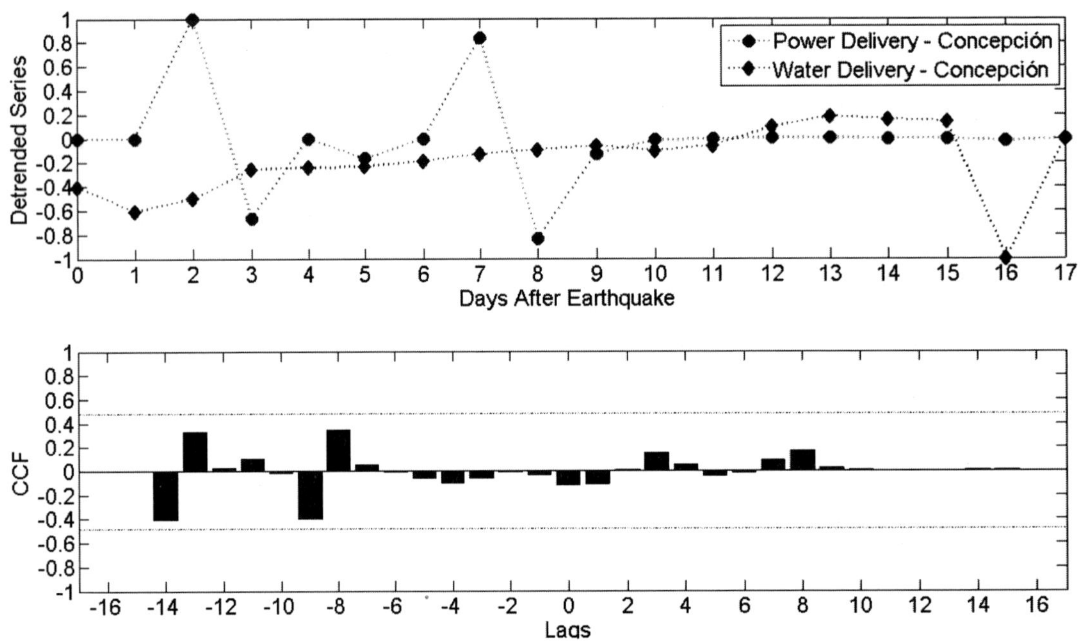

Figure A.17. Detrended series and cross-correlation CCF values between power in Concepción and water in Concepción.

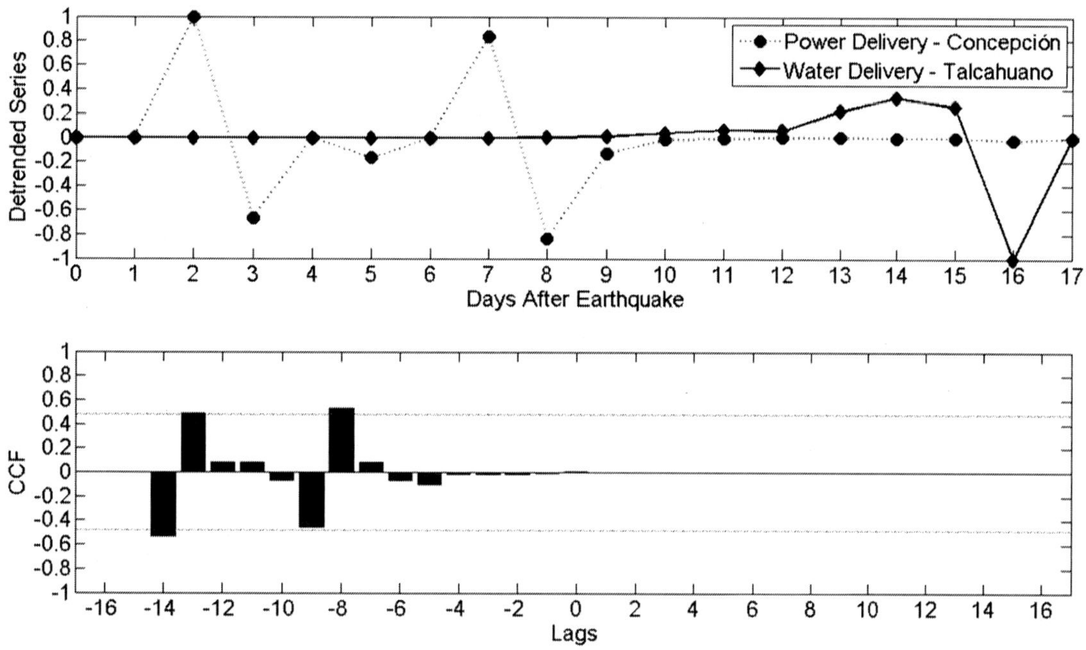

Figure A.18. Detrended series and cross-correlation CCF values between power in Concepción and water in Talcahuano.

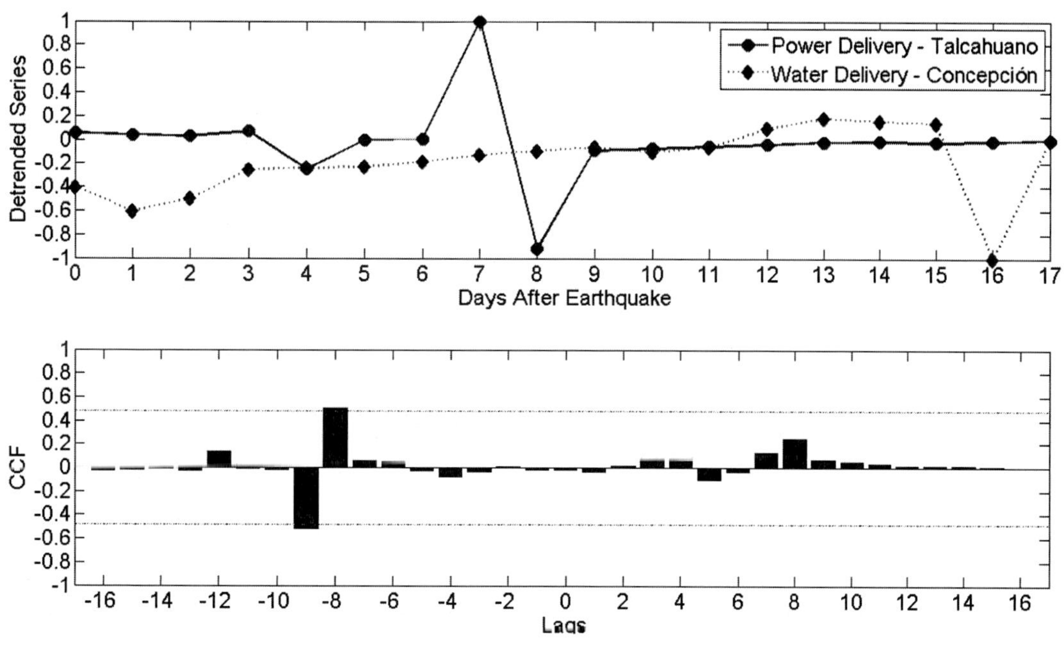

Figure A.19. Detrended series and cross-correlation CCF values between power in Talcahuano and water in Concepción.

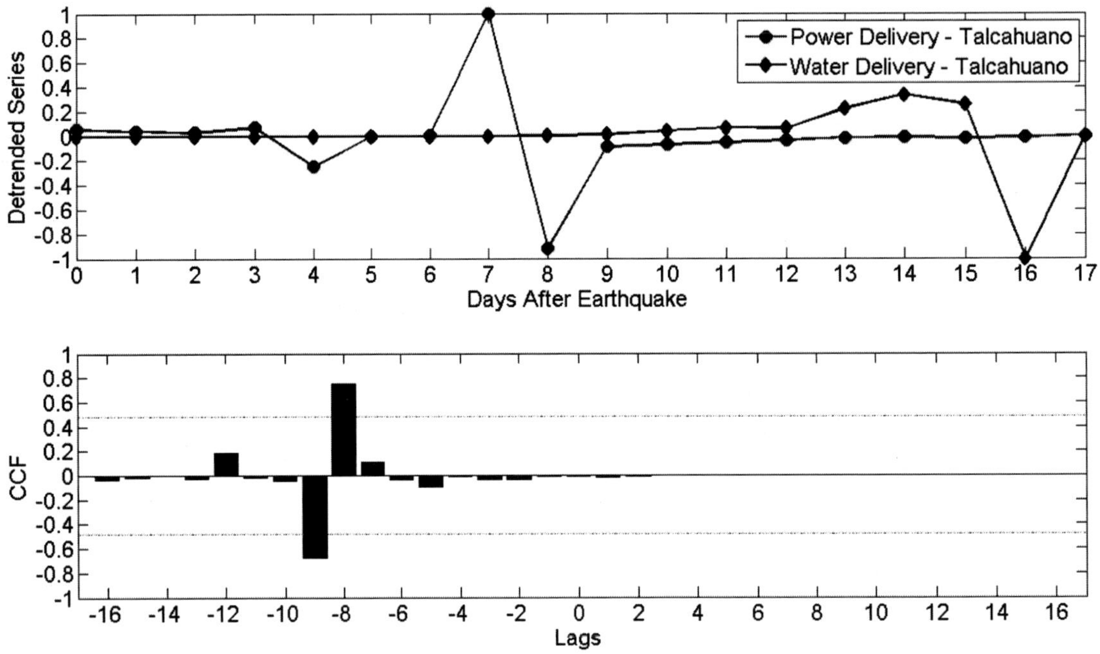

Figure A.20. Detrended series and cross-correlation CCF values between power in Talcahuano and water in Talcahuano.

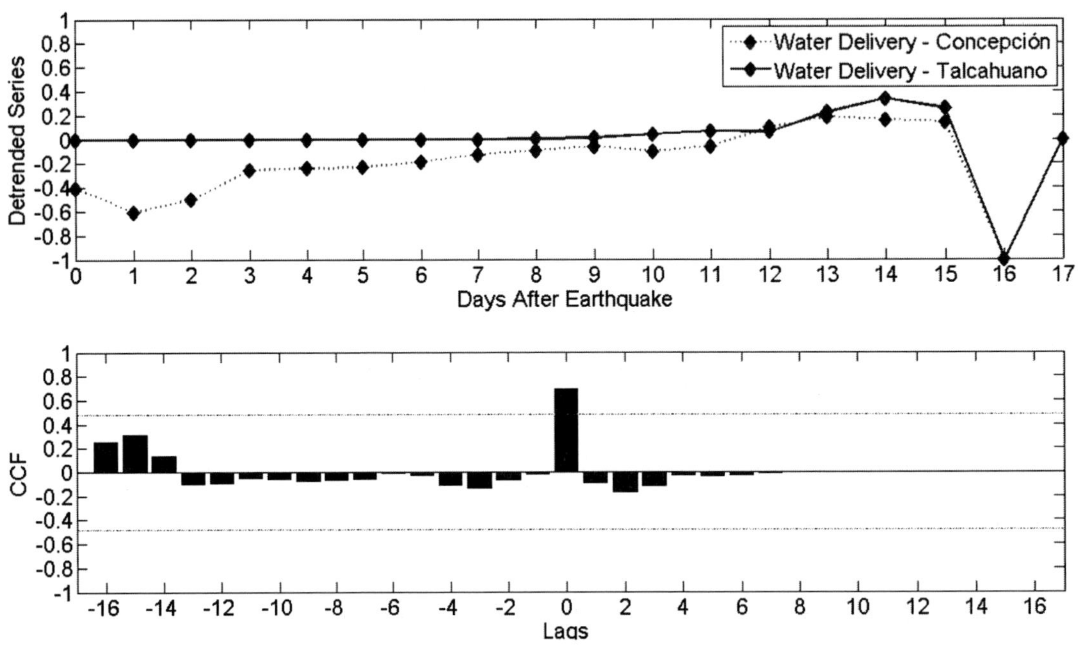

Figure A.21. Detrended series and cross-correlation CCF values between water in Concepción and water in Talcahuano.

11 EMERGENCY RESPONSE, RECOVERY, AND SOCIAL IMPACT

The February 27, 2010 Maule Earthquake magnitude 8.8 earthquake and tsunami struck the central region of Chile. While the country's capital Santiago suffered some damage, the greatest impacts were felt in the cities of Concepción, Talcahuano, and other central valley towns and coastal towns within the rupture zone. The total fatalities resulting from the earthquake and tsunami stood at 521 deaths; over half were from the tsunami at the time of this report. Many of the casualties related to the tsunami occurred near Constitución in campgrounds along and on islands in the River Maule where refuge to high ground was not possible.

The earthquake and tsunami reportedly damaged over 370,000 homes. According to the Ministry of Health, 17 medical establishments required reconstruction, and 54 required repairs, effectively taking 5,000 beds out of service and jeopardizing 10,000 more. Highways, airports, ports, and urban and rural infrastructure as well as other lifelines such as the electrical systems, communication systems, water and wastewater systems, refineries, and liquid fuel and natural gas distributions system were all significantly affected.

However in general, Chile recovered reasonably well from an earthquake of this magnitude. Despite significant damage to residential and commercial adobe structures, extensive tsunami damage, and many severely damaged newer buildings and bridges, Chile suffered minimal deaths for an earthquake of this size and duration. In addition, with the exception of areas of adobe construction and areas hit by the tsunami, Chile showed great resiliency during recovery. Ruta 5 was passable for the length of the country within days, and the majority of utilities were in operation within a week. However, at the time of the TCLEE reconnaissance, adobe and tsunami areas were completely destroyed, mobility in Concepción was severely impaired by bridge damage, and the fishing industry was severely disabled by uplift and loss of infrastructure.

Table 11.1 provides an early estimate in U.S. dollars of the infrastructure damage, as assessed by the Chilean Chamber of Construction Companies. Several months after the earthquake, the Chilean government projected the damage at about $30 billion, which is 18 percent of Chile's gross domestic product. Additional socioeconomic information can be found in reports at *www.eeri.org*.

Table 11.1. Estimate of Loss (Source: Data from Balance de la Infraestructura en Chile Camara Chilena de la Construccion Santiago, June 2010)

Summary (USD Millions)			
SECTOR	Total Investment Required	Public	Private
Highways	706	526	180
Airports	21	1	20
Ports	470	470	0
Hospitals	2700	2700	0
Prisons	80	80	0
Potable Water (Rural)	47	47	0
Potable Water (Urban)	226	0	226
Hydraulic Structures	50	50	0
Power	360	0	360
Total	**4660**	**3874**	**786**
Percentage	**100%**	**83%**	**17%**

11.1 Earthquake Culture

Chileans as a group are well equipped to survive the initial dangers of an earthquake as well as the subsequent tsunami. As discussed in Chapter 2, the Nazca Plate along the Chilean coast is subducting under the South American Plate at a rate of 5 to 8 cm/yr. This results in a very seismically active area. Chile has had six earthquakes exceeding magnitude 8 since 1900, and smaller quakes are very common. As a result, the population is well educated through experience, years of preparation in school, and public awareness campaigns. These campaigns cover what to expect during an earthquake, how to survive the shaking, and the critical importance of immediate evacuation in low-lying coastal areas. This knowledge resulted in minimal deaths for an earthquake of this magnitude. The population was prepared to escape harm in collapsible adobe structures and from tsunami waves.

A typical response is that of the chief air traffic controller in Carriel Sur, who drove his family to the airport parking lot. He parked with the car pointed away from the sea as a precaution against a possible tsunami, as he entered the building to assess the damage.

11.2 Emergency Response

Emergency response after the earthquake could be improved in several ways. In general, the problems stemmed from a changing political climate, failures in communication, and the lack of a cohesive national earthquake program. The TCLEE team spoke with government officials, local engineers, and the public to learn about the response immediately after the earthquake, the problems encountered, and the public's reaction to the response.

Government Response

The disaster could not have come at a more vulnerable time for the government. The outgoing Bachelet administration, while still formally in power awaiting the official inauguration of president-elect Pinera, was undermanned with key officials on vacation anticipating their return to the private sector. This appeared to hamper initial government response, at least in Region 8, where the position of Intendente, the "appointed governor" responsible for leading the disaster relief effort, was vacant.

As government relief failed to materialize in the first days following the earthquake, citizens of the hardest hit towns began to search for food in nearby warehouses. In some cases, this activity was covered on television internationally as desperate looting in lieu of an effective government response. At this point President Bachelet declared martial law. The armed forces then took a principal role in the relief effort, deploying civil-military teams to the most heavily damaged areas to handle emergency response, rescue, and relief.

The delivery of emergency supplies was affected by significant damage to the north-south highways and the temporary closure of the Santiago International Airport. The airport was closed to departures due to suspension of fuelling services while the in-ground hydrant system underwent exhaustive inspections. As a consequence, relief was initially brought to the Concepción region via Navy ships. With most of the ports damaged, supplies were transferred via helicopter to the affected areas. This initial relief effort lasted for approximately 1 week, until supplies could be transported effectively by road.

The military provided victims potable water stored in bladders (Figure 11.1) using reverse osmosis (RO) process.

During the week of the TCLEE site visit, the military formally returned administration of the region to the Intendente, marking the end of the relief effort and the beginning of restoration.

One of the most memorable images from the media coverage of the Maule quake was the videos of looting of stores, port facilities, and fish factories that occurred in Concepción and the tsunami damaged areas of Talcahuano. It was evident during our conversations with the public and various officials that communications from the government about emergency procedures, facilities, and the state of the country were severely lacking in the first 24 to 48 hours after the earthquake, especially in the hardest hit areas. Assurances that emergency aid and personnel to help were available and on the way, as well as information about aftershocks, personal safety, and the tsunami were inadequate to ease public anxiety. The military deployment to areas where looting and panic were rampant occurred significantly later than most thought necessary.

However, possibly the most important failure in emergency response occurred immediately after the earthquake. Our conversations indicate that there was a misunderstanding about the nature of the earthquake and the possibility of a tsunami. The Chilean Navy's Department of Hydrology and Oceanography (Servicio Hidrografico y

Figure 11.1. Potable water bladder

Oceanographico de la Armada de Chile, or SHOA) mistakenly thought that there would be no tsunami because the epicenter of the earthquake was below the continent. This information was passed on to the president, and within an hour of the earthquake while tsunami waves were reaching the shore, the President announced over national radio that a tsunami would not occur. Thankfully, the population was experienced enough with subduction zone earthquakes and the resulting tsunamis to disregard the warning and stayed clear of the tsunami. However, the mistake was grave enough that it resulted in the resignation of a high-ranking government official within the decision chain. As discussed in Chapter 2, the rupture zone of the quake was very large and the surface rupture of the fault was under the ocean, miles from the true epicenter. This misunderstanding could have been avoided if Chile had installed an integrated national earthquake program including geologists, SHOA, and the national emergency management agency, with staff geologists to interpret data and advise the government during seismic events. In a location as seismically active as Chile, this should be a national priority.

International Response

Soon after the news of the Maule earthquake was announced to the world, search and rescue aid, financial support, and donation of necessities poured in from many countries. Carriel Sur, the regional airport, never closed even though the Instrument Landing System was out of service for a few days due to concerns that the earth movement had affected the precision of the localizer and ILS antenna. A certification flight from Santiago reestablished full operations.

Relief flights eventually were able to reach the regional airport of Carril Sur, with operations peaking at 350 per day—a sevenfold increase over its normal operations. The air traffic control facility, operating without radar control, was expanded from two to six controllers, directed by a shift supervisor. Controllers slept in the break room when relieved. This activity lasted for almost 3 weeks, until replacement personnel arrived.

11.3 Recovery

With the exception of tsunami zones and areas with adobe construction, the recovery of basic utilities and transportation was relatively quick. Immediately after the earthquake, Santiago was initially isolated from the central and southern regions of the country, with damage to the main artery Ruta 5 due to bridge and embankment failures in distinct points from Puente Maipo to Rio Bueno. The Santiago-Talca stretch, operated by the Autopista del Maipo concessionaire, required alternative routes due to major damage to the roadway and overpasses. Despite the damage, Ruta 5 was generally operational within 5 days, once overpasses were shored, embankments repaired, and temporary routes established. Water, electricity, and phone systems were also generally operational within a week in all but the hardest hit areas.

At the main airport in Santiago, air traffic controllers on duty were able to evacuate the main ATC tower, which suffered structural damage, carrying portable communications equipment to transfer operations to an older, back-up tower that served one of the runways. This permitted some arriving flights to land immediately after the earthquake. However, operations remained severely restricted for almost a week, followed by a rapid recovery. The annual Chilean air show, FIDAE, was held successfully on both runways just 3 weeks after the earthquake.

However, in the adobe and tsunami zones where the majority of the towns and residences were completely demolished or washed away, it will take months if not years to rebuild. Large tent cities developed near demolished coastal towns. At the time of the TCLEE reconnaissance, displaced residents in Talcahuano were still living in tent cities, as shown in Figure 11.2.

The military had begun construction of more substantial temporary housing for the displaced by 4 weeks after the quake (Fig. 11.3). Temporary housing consisted of approximately 10 by 12 ft. wooden houses with electricity (Fig, 11.4 and 11.5). However, during the TCLEE reconnaissance 5 weeks after the initial quake, construction had just begun, and the demand for housing was growing due to the coming winter and wet season.

In coastal areas such as Lebu, where uplift rendered the harbors useless, and at ports such as San Vincente and Coronel, where fishing piers were severely damages, one of Chile's largest industries was severely impaired. If repair of uplifted ports is possible, it will take months to reestablish channels for fishing vessels. Reconstruction or relocation of fishing piers also will require months. In the meantime, the many Chileans who are dependent on the fishing ports to survive have relocated to areas that have working facilities.

Figure 11.2. Tent city in Talcahuano

Figure 11.3. Chilean Military constructing temporary house for the victims in Dichato.

Figure 11.4. Temporary housing units in Dichato

Figure 11.5. The tent city in Talcahuano

Concepción is a unique coastal city in that it is split by the Bio Bio River. The majority of the city's approximately 212,000 residents and an additional 675,000 Chileans residing in the surrounding towns live on the south side of the river. However, the main commercial areas, airport, and industry are on the north side. All three of the main bridges across the river failed during the earthquake. The department of public works was able to get one lane in each direction across the Llacolyn Bridge open within a week, but six weeks after the quake the city was still working with driving restrictions. As a result, public demonstrations were held to protest the slow recovery of the bridges.

Private Industry Response

The reconstruction of infrastructure destroyed by the earthquake/tsunami will fall on both the public and private sectors. The Chilean Chamber of Construction Companies estimates that $4.6 billion is required for infrastructure, of which 17 percent is private and the rest public. The projects to be addressed with private funds include the highways of Santiago and Ruta 5, the airport, potable water, power, and ports.

For example, most ports in Region 8, which handles 55 percent of the country's maritime cargo (about 100 million tons/yr.), are in private hands. These ports are for the most part highly competitive and attractive to foreign trade and, with the notable exception of Talcahuano, quickly returned to full operations. Their performance in the first few days after the crisis at the Santiago International Airport, which is also in private hands, has raised questions, however. Business News Americas reported that the Ministry of Public Works is planning a new airport in Tongoy, 400 kilometers to the north of Santiago. In addition to helping develop this region, this facility will back up the Santiago terminal in case of emergencies.

11.4 Social Impact

As prepared as Chileans were for surviving an earthquake, shaking of this magnitude and intensity and damage on this scale greatly affected the overall public well-being. Further, the loss of critical services, including water, wastewater, electricity, gas, and phones, seemed to produce public shock. The failure of these critical lifelines in the hardest hit areas caused most residents to stay home with their families for the first week. Even contractors seemed immobilized by the event.

The impact on the public was especially great in the areas devastated by extensive collapse of adobe structures or tsunami waves. Six weeks after the quake, the rubble of whole towns was still being removed, rough temporary housing was scarce and still under construction, and people who had lost everything, including their livelihoods, were still wandering the area trying to recover belongings from the rubble. In the island campground area of Constitución where more than 100 tourists were killed by tsunami waves, military personnel were still recovering bodies.

Coastal towns not destroyed by the tsunamis were cut off from the remainder of the country due to road and bridge failures. Towns dependent on tourist income lost their livelihoods. Towns dependent on fishing ports that were damaged were also heavily affected. Many fishermen have already moved to ports farther away to unload their catches and continue their livelihoods. Entire towns will have to be rebuilt or relocated to safer ground.

In areas with adobe structures and other areas with significant structural damage, the government evaluated buildings and marked those scheduled for demolition with spray paint. In these areas, thousands of people were displaced and commercial operations were still closed more than 6 weeks after the earthquake. Many areas did not have electricity or water other than emergency stations. Signs asking for food and water were posted in the hardest hit rural areas, along roadways, and where displaced people gathered and felt safe.

11.5 Major Observations and Recommendations

There was a short outbreak of looting in Concepción, which was handled by the Chilean military without any incident within a couple of days. During our visit about 5 weeks later, there was no presence of military in Concepción. Depots were set up a week after the looting for people to return stolen merchandise.

The waterfront in Talcahuano, which had extensive damage from both the earthquake and tsunami, was guarded by police to prevent unauthorized access.

The loss of many churches in the area reduced activities that bond community members together to help each other, particularly after a disaster like this. Chileans should look into upgrading their churches so they can continue to function as a social gathering place and shelter during emergencies.

As discussed in Chapter 2, the TCLEE team recommends that Chile develop an integrated national earthquake program, such as the one the country has developed for volcanos. The program should incorporate the national geologic service, Sernageomin; the national emergency management service; and SHOA, the national oceanic society, to prevent mistakes such as the false all-clear tsunami announcement.

Further, cell service was interrupted within hours due to the short duration of battery backup and equipment anchoring at cell sites, which impaired communication among the public and within government agencies. Wireless service providers should consider increasing the battery life of cell sites to reduce the loss of service after power failure or develop a better plan for emergency recovery structures to deal with service disruptions.

11.6 Acknowledgements

Many people provided information and assistance with this chapter, including SNGM, numerous Chilean contacts, and TCLEE team members. We also would like to thank Prof. Mauricio Villagran, Ms. Karla Gonzalez of GHD Consulting, and Mr. Geri Pranzini of GHD Consulting. Special thanks go to concerned Santiago citizens Ms. Barbara Domont, and Mr. Nicolas Lira Turpaud.

Unless otherwise noted, all figures and tables belong to ASCE/TCLEE team members.

Index

Page numbers followed by *f* and *t* indicate figures and tables, respectively.